The Best American Science Writing 2003

The Best American Science Writing

The Best American
SCIENCE WRITING

2003

EDITOR: OLIVER SACKS

Series Editor: Jesse Cohen

ecco

An Imprint of HarperCollinsPublishers

THE BEST AMERICAN SCIENCE WRITING

Permissions appear following page 269.

Compilation copyright © 2003 by HarperCollins Publishers.

Introduction copyright © 2003 by Oliver Sacks.

HarperCollins books may be purchased for educational, business, or sales promotional use. For information, please write: Special Markets Department, HarperCollins Publishers Inc., 10 East 53rd Street, New York, NY 10022.

FIRST EDITION

Designed by Cassandra J. Pappas

Library of Congress Cataloging-in-Publication Data is available upon request.

ISBN 0-06-621163-8 HARDCOVER
ISBN 0-06-093651-7 TRADE PAPERBACK

03   04   05   06   07   BVG/RRD   10   9   8   7   6   5   4   3   2   1

Contents

Introduction by Oliver Sacks

Science writing—unpleasant term, but what else should one call it?—first presented itself to me, as a boy, in books. I was given *The Stars in Their Courses*, by the famous astronomer James Jeans, when I was ten, and this so excited me that I then wanted to read everything else he had written. I rushed to the local public library (much of my real education came from reading in public libraries, rather than from lessons at school), and went straight to J. I took out *The Mysterious Universe* and devoured it, then *The Universe Around Us* and *Through Space and Time*.

Through Space and Time was based on the Christmas Lectures that Jeans delivered at the Royal Institution in London in 1933 (this date particularly appealed to me, since it was the year of my birth). Such lectures, designed for general audiences, had been an important feature of the RI's activities since its founding in 1799. The early nineteenth century marked, in many ways, the beginnings of modern science (the very word "scientist" was coined only in the 1830s; there had just been "savants" and "natural philosophers" before), and from the start, there was a need to present the latest discoveries in exciting, accessible terms. Humphry Davy, the great poet-chemist who discovered the alkali metals, and his student Michael Faraday, who went on to discover many fundamentals of chemistry and electricity, both lectured at the Royal Institution, where they attracted huge crowds and became a major part of cultural life in London; and they did much to fuel what was to become a great popular interest in science.

Presenting new scientific concepts and discoveries in an accessible and attractive form was not regarded as a contemptible activity, or as a distraction from the actual, serious work of science. Thus between 1835 and 1860, Faraday delivered five or more Evening Discourses every year, covering an enormous range of topics, and he obviously enjoyed giving these. This sense of delight in science shines through all of his lectures and writings (his *Chemical History of a Candle*, based on his famous Christmas Lectures at the RI, is still as delightful and stimulating today as when it was first published in 1861).

By the middle of the nineteenth century, the general public, confronted by the rapidly changing technology of the industrial revolution no less than the revolution in understanding that was happening in chemistry, natural history, and biology, needed a way for such scientific information to be widely disseminated, and in terms the layman could readily comprehend. It was during this period that the great museums of natural history and of science were established throughout America and Europe, and that popular science magazines came into being. *Scientific American*, founded in the 1840s, was one of the first and most distinguished of these, and it continues to be published regularly every month, as it has for more than a hundred and fifty years. Indeed, until the middle of the twentieth century, one could read *Scientific American* and stay reasonably current with scientific development in general. But the last thirty years or so have seen so huge an explosion in scientific knowledge, the creation of so many new disciplines and subdisciplines, that it is no longer possible for a single person to keep current in all fields of science, and it is difficult for any but the most specialized to read most scientific journals.

Perhaps for this very reason, the popular appetite for good science writing has increased too, and the venerable *Scientific American* has been joined by an ever-increasing number of other magazines aimed at the nonspecialist, some concerned exclusively with science, or with particular branches of science, and others with occasional articles on science.

I do, I confess, voraciously read such magazines and periodicals, partly to keep up in my own field, but also to learn what is going on, what is being discovered and thought in fields far from my own, but also for the excellence of writing they often contain. I subscribe to many specialized periodicals: *Neurology, Brain*, and so on, because I am a neurologist; *Pteridologist* and its lighter cousin, *Fiddlehead Forum*, because I am a fern lover; *Mineralogical Record*, because I love minerals; *Journal of the History of Neurosciences* and *Ambix*, because I am fascinated by the history of science. But I subscribe to a clutch of popular ones as well. Having just returned to my desk after a couple of weeks away, I find a massive accumulation of these—*Science News, Scientific American, New Scientist, Discover, National Geographic, American Scientist* (to say nothing of *Science* and *Nature, The New Yorker, The New York Review of Books, Harper's Magazine,* and *The Atlantic Monthly*).

I have stuffed them all into my briefcase, wondering if I will tear the handle off with their weight, and after dinner I will lie down on my bed (I do most of my reading there, or in the bath), and rush through them eagerly and greedily, picking out (sometimes razoring out) articles that stimulate me because of new ideas or information. An omnivore, yet selective, a sort of filter-feeder, I will extract intellectual nutrients from the articles as I extract nutrients from my dinner. Every so often, however, I am arrested by an article because it contains not just new information but a highly individual point of view, a personal perspective, a voice that compels my interest, raising what would otherwise be a report or a review to the level of an essay marked by clarity, individuality, and beauty of writing.

Reading an article by the late great Stephen Jay Gould always gave me this special sense—one felt the man, his special interests and experiences (whether his favorite snail, *Cerion,* punctuated equilibrium, or Gilbert and Sullivan), and the landscape of his mind, whatever the particular theme of the article; and one felt the highly individual choice of image and language. A Stephen Jay Gould article was never predictable, never dry, could not be imitated or mistaken for anyone else's.

I am not entirely sure what makes "good" science writing (or indeed, "good" writing of any sort), but Coleridge put it as succinctly as possible when he advised "proper words, in proper places." The best science writing, it seems to me, has a swiftness and naturalness, a transparency and clarity, not clogged with pretentiousness or literary artifice. The science writer gives himself or herself to the subject completely, does not intrude on it in an annoying or impertinent way, and yet gives a personal warmth and perspective to every word. Science writing cannot be completely "objective"—how can it be, when science itself is so human an activity?—but it is never self-indulgently subjective either. It is, at best, a wonderful fusion, as factual as a news report, as imaginative as a novel. It is with this in mind that I have made what is bound to be a highly partial and idiosyncratic selection of the best American science writing of 2002. I have, frankly, found this to be nearly a impossible task—for I would like to have included twenty or a hundred pieces for every one here, and to have given every facet of science, from paleontology to psychoanalysis, its place and due. But there is only so much one can do in a book of three hundred pages.

Science writing, good science writing, is not confined to "scientific" magazines (though this collection, not surprisingly, includes articles that appeared in *Discover, Scientific American,* and *Popular Science*). It is equally to be found in general publications, such as *The Washington Post, The New York Times, The New Yorker, The New York Review of Books, Harper's Magazine, The Atlantic Monthly,* and *The New Republic*; as well as many less widely circulated but highly regarded magazines such as *Daedalus, Monthly Review, Mother Jones, Tikkun,* and *Southwest Review*; and even local and regional publications such as *LA Weekly* and *High Country News*. All of these are represented in this collection, as well as an online magazine, Salon.com, and *Wings,* the tiny journal of the Xerces Society, dedicated to the conservation of invertebrates. A very slender magazine—only a dozen pages or so—that comes out every week is the admirable *Science News,* which in addition to scientific news always contains original articles and essays, and I am glad to be able to include a piece from this publication.

The rules of this series prevent the inclusion of any articles from non-American authors, and thus this volume does not contain anything from two great periodicals published in England (but freely available in the United States)—*New Scientist,* an extremely fine popular science weekly with no exact analogue in the United States, and *Nature,* which not only represents a world forum for original scientific articles (it is just fifty years since Watson and Crick published their famous letter in it suggesting the structure of DNA), but also contains some of the very best science writing one is likely to see. It is similar to its American counterpart, *Science.*

Though there have always been scientists who have excelled (and delighted) in lucid expositions of their own and others' work, and there has always been coverage of major scientific discoveries in the general press, "science writing" as such is a relatively new phenomenon, yet it is one that has already achieved a central place in our culture, as this series attests. There are now dozens of first-class science writers whose names are well known to every reader, and whom one can always turn to with the near certainty of encountering clarity, enthusiasm, and depth. There is the temptation, in an anthology such as this, to rely on these tried-and-true names, but I have tried here to introduce new writers as well, whose names may now be unfamiliar but will not be for long.

At the opposite end of the spectrum, Stephen Jay Gould was, unarguably, the best known and most beloved science writer of the past quarter century. As I write this, it is not quite a year since Steve died so prematurely, at the age of sixty, in May 2002. Steve was everything—a field scientist, a theoretician, a historian of science, a bibliophile; but he was also an unabashed lover of oratorios, baseball, and old buildings. The miracle of his writing was that he brought everything together, brought the whole of himself into his writing. His subjects sometimes seemed recondite or odd—he had a special feeling for the overlooked, the unconsidered, the forgotten, the dismissed—but by the time he had dealt with them, they seemed the most interesting things in the world, and the world seemed richer for having them restored to it. And so it seems fitting to dedicate this volume, celebrating our best science writing, to his memory.

The Best American Science Writing 2003

PETER CANBY

The Forest Primeval

FROM *HARPER'S MAGAZINE*

> *Nouabalé-Ndoki is one of the most remote places on Earth—seventeen hundred square miles of nature preserve in a hard-to-reach region of the Republic of Congo. Stephen Blake, an English zoologist for whom the word "intrepid" seems an understatement, has made several monthlong journeys, on foot, through Nouabalé-Ndoki—undeterred by lack of modern amenities, the threat of disease, and the presence of poisonous snakes—to track its remarkable population of elephants. The writer Peter Canby accompanies Blake on what is to be his last trip and observes a scientist at home at the edge of the world.*

I've just reached Makao, the most remote village in the Republic of Congo. I'm traveling with Stephen Blake, a British wildlife biologist, in a thirty-foot, outboard motor–powered pirogue—a dugout canoe—following the muddy, weed-clotted Motaba River north from its confluence with the Ubangui River. At first, after leaving the Ubangui, we passed small villages hacked out of the forest, but for a long time we've seen swamp interrupted only by the odd fishing camp: small bird nest–like huts and topless Pygmy women in grass skirts waving their catch forlornly as we motor by.

But now we've arrived at Makao, the end of the line, the last town along the Motaba. Ahead is pure, howling wilderness. Makao has a population of perhaps 500, half Bantu and half Bayaka—among the most traditional Pygmy tribes in Africa. The village long had a reputation as a poaching town, one of the centers of the extensive and illegal African "bushmeat" trade, which, in the Congo basin alone, still accounts, annually, for a million metric tons of meat from animals that have been illegally killed. But since 1993 the poaching in Makao has all but ceased, and the village has taken on another significance: it is the back door to the Nouabalé-Ndoki forest. Nouabalé-Ndoki is named for two rivers, only one of which actually exists. The name of the existing river—Ndoki—means "sorcerer" in Lingala, the lingua franca of much of the two Congos. Nouabalé doesn't mean a thing. It's a misnomer for another river, the Mabale, inaccurately represented on a geographer's map in the faraway Congolese capital, Brazzaville.

Nouabalé-Ndoki is now a 1,700-square-mile national park known chiefly for having the least disturbed population of forest life in Central Africa. No one lives in the park, or anywhere nearby. Nouabalé-Ndoki has neither roads nor footpaths. It contains forest elephants, western lowland gorillas, leopards, chimpanzees, forest and red river hogs, dwarf and slender-snouted crocodiles, innumerable kinds of monkeys, and nine species of forest antelope, including the reclusive sitatunga and the supremely beautiful bongo. The southwest corner of the park is home to the famous "naive chimps" that sit for hours and stare at human intruders. Until biologists arrived just over ten years ago, few of these animals, including the chimps, had ever encountered humans.

Blake studies elephants. A self-proclaimed "working-class lad" from Dartford, England, Blake read zoology at the University of London; he is now working on a doctoral thesis about the migratory patterns of Nouabalé-Ndoki forest elephants at the University of Edinburgh. Thirty-six, fit, and lean, Blake is known as a scientist who likes the bush and is not afraid to go where wild animals live. But he's also considered audacious, a biologist who thinks nothing of crossing wild forests clad in sandals and a pair of shorts. Richard Ruggiero, who runs the elephant fund for the U.S. Fish and Wildlife Service and worked with Blake just after the park was established, compares him to nineteenth-century explorers: "He's someone who could walk across Africa, turn around, and then be ready to go back again." Another colleague described encountering him as he emerged from a long stint in the bush. "He was wearing torn shorts and a tattered T-shirt. He had a staph infection but seemed completely happy."

As part of his research, Blake has taken a series of what he calls "long walks"—foot surveys that start in Makao and follow a web of elephant trails up the Motaba and Mokala rivers to the park's northern border, cross the park from north to south, and then emerge from the headwater swamps of the Likouala aux Herbes River below the park's southern border. (The gorillas of the Likouala aux Herbes were the subject of Blake's master's thesis at Edinburgh.) Each of these treks—and Blake has made eight—covers about 150 miles and takes about a month. When I joined him, Blake was preparing to embark on his ninth and final trip along his survey route. I had heard of Blake's work from Amy Vedder, a program director at the Wildlife Conservation Society, which, along with the U.S. Fish and Wildlife Service and the Columbus (Ohio) Zoo, funds his research. Vedder and I had been discussing the toll that the region's wars have taken on its wildlife when she told me about Blake's long walks. I signed on to accompany him on his last one. At the time, it seemed a rare opportunity to see the Earth as it was thousands of years ago, at the moment when humans lived side by side with the great apes from which they evolved.

But now that I've reached Makao, I'm wondering why I made no special preparations for this trip. All the perils, which seemed theoretical before I left, have become disturbingly real. Not only don't we have phones or any means of communication; we also face threats of dengue fever, deadly malaria, the newly resurgent sleeping sickness, and even AIDS and Ebola, which are believed to have emerged from the forests of this region. I'm also afraid of army ants, ticks (eventually one crawls up my nose and inflates just at the top of my nasal passage), swarms of flies, and, above all, snakes. When I let slip that I am particularly nervous about snakes, Blake tells me about the Gabon viper, a fat, deadly-poisonous snake with the longest fangs of any snake in the world. It often lies in ambush on Nouabalé-Ndoki trails. "The Gabon viper always bites the third person in line," Blake says glibly. "That's your slot."

THE WILDLIFE CONSERVATION Society maintains a field station in Makao, and we spend several days there assembling a crew. One morning, as Blake and I bathe in the Motaba while a cloud of blue butterflies swarms around us, he explains how his recruiting policy has been determined by local economics. Bushmeat, he tells me, was a staple of the Congolese diet and, for many, the only available source of income. In Makao, the WCS provides jobs to people who are now forbidden by law to hunt; Blake himself has also sought to hire the best former hunters in order to keep them off the market. Practically speaking, this means recruiting the Bayaka, who live not just in Makao but also north and east of the park. Unlike Pygmies elsewhere in Africa, who are increasingly removed from hunting and gathering, many of the Bayaka still go into the forest for months, or even years, at a time, living off the land with little more than spears and homemade crossbows. Blake hires them because they know the forest intimately. "I often think every Bayaka should be awarded a doctorate in forest ecology," Blake says. "They know what's going on."

But Makao is ruled by Bantus, who, while dominant, know much less than the Bayaka about the forest. Blake would rather travel only with Bayaka, but, because of the dynamics of the village, he also hires Bantus. The relationship between the groups is complicated. The Bayaka Pygmies are small forest people—the men in Makao seem to average around five feet three—and presumably the original inhabitants of Central Africa. The Bantus, who are taller, are fishermen and slash-and-burn cultivators who migrated to the region several thousand years ago. The Bantus control Bayaka families; the Bayaka are expected to hunt for their Bantu owners and to work their manioc fields. In return the Bayaka get metal implements, notably cooking pots and spear points, made from automobile leaf springs; having acquired these things, they light off to follow a nomadic life in the forest. The arrangement is changing, however, as many Bayaka now live in the village year-round. Not all of the Bayaka still know how to make crossbows, recognize plants, or use spears. They can no longer survive in the forest.

Several of Blake's Bayaka recruits have accompanied him on earlier treks. They include one of Blake's oldest Bayaka friends, Lamba, who is named for a stout vine that winds helix-like up into the canopy trees, and Mossimbo, who is named for an elephant-hunting charm. But this time Blake is excited about a new recruit: Zonmiputu. Zonmiputu comes from one of the most traditional bands of the Makao Bayaka. Blake had met him on one of his early trips after a chance encounter, somewhere outside the park, with Zonmiputu's father's band, which had been living off the forest, following the ancient, intricate Bayaka way of life, for more than a year.

"They were carrying spears and homemade crossbows," Blake recalls. "They had one cooking pot, no water jugs, and a lot of baskets they'd made out of forest vines. Their clothes had worn out, and they'd gone back to wearing bark fabric."

As the first person ever to have employed the Bayaka, Blake is changing their lives. "Before they worked for me, their wives had to scrape for yams using sticks. Almost all their food was baked in leaves. Now one of them works for me for a month and makes enough money to buy a machete, a few clothes, a pot, and some fishhooks." Still, after returning from a month in the forest, Blake has frequently been confronted by Bantu *patrons* demanding the money he is about to pay "their" Pygmy. They react with incredulity when Blake won't give it to them.

As Blake and I talk by the river, I hear what I take for a birdcall. It's soon answered by a similar call—but at a harmonic interval—and then a third. Soon the river valley is full of strange syncopated harmonies. It's as if the trees themselves were singing. "Pygmies," Blake says when he sees my puzzled expression. "They're working the fields."

BY THE NEXT DAY we've assembled our team—Zonmiputu, Lamba, Mossimbo, four other Bayaka, three Bantus, Blake, and me. Our walk begins another six hours up the river. We pile ourselves and our gear into the pirogue. Our "tucker," as Blake calls our food, comes from a market in the town of Impfondo along the Ubangui. It consists of sixty cans of tomato paste, two hundred cans of Moroccan sardines, forty cans of Argentine corned beef, twenty pounds of spaghetti, one hundred pounds of rice, several bags of "pili-pili"— the very hot, powdered African peppers—and large quantities of cooking oil, sugar, coffee, and tea. ("What's an Englishman to do in the forest without tea?" Blake asks.) We've topped off our supplies with three fifty-pound sacks of manioc flour and two baskets of smoked Ubangui River fish, bought from a fish merchant in an Impfondo courtyard.

We cast off early one morning. Above Makao, the riverbanks are uninhabited. It's late February—the end of the dry season—but the twenty-foot-wide river courses swiftly between marshy banks. We pass African fish eagles, perched on overhanging branches. Hornbills wing their way overhead, making otherworldly cries and beating the air with a ferocity that evokes the original archaeopteryx. Around ten in the morning an eight-foot, slender-snouted crocodile surfaces next to the boat and glances dispassionately at us. Our disembarkation point, from which the boatman will return the pirogue to Makao and we will begin walking, is near a fallen tree just below the juncture of the Motaba with one of its tributaries, the Mokala. I step ashore, look down at my pale, tender feet clad in rubber sandals, and wonder how I'm going to survive this expedition. In front of me, hearts of palm have been peeled—evidence of gorillas. Behind me Mossimbo spots fresh python skin, assumes the python is nearby, and leaps back in panic. Pythons here can grow to twenty feet; they strangle everything from antelopes to crocodiles. Everyone roars with laughter at Mossimbo's expense. The laughter covers the whir of the pirogue's motor as it pulls away, and when the Pygmies quiet down I hear the pirogue disappearing back downriver. My heart sinks.

———

TEN YEARS AGO the Nouabalé-Ndoki park didn't exist. The land was set aside after a decade of mass slaughter of elephants. During the 1970s a Japanese vogue for ivory signature seals, a consequent tenfold increase in the price of ivory, and a continent-wide collapse of civil authority combined to set off an orgy of elephant destruction. Poachers wielding AK-47s massacred entire herds for tusks, and then sold the ivory through illegal networks presided over by potenates like Jean-Bédel Bokassa, the cannibal emperor of the Central African Republic, and Jonas Savimbi, the murderous Angolan warlord. At the height of the slaughter, poachers were killing 80,000 elephants annually. In the 1980s almost 700,000 elephants were killed.

In 1989 conservation organizations intervened. The Convention on International Trade in Endangered Species (CITES), a widely supported treaty that regulates trade in endangered species, put African elephant ivory on its list of most restricted commodities, thus effectively banning its international exchange. The market collapsed and conservationists rallied to save the remaining elephants. Africa has two types of elephants: *Loxodonta africana africana,* the bush elephant of the savannas, and *Loxodonta africana cyclotis,* the forest elephant. Biologists know a great deal about the savanna elephant, the world's largest land mammal, which is easy to spot and easy to monitor. But the forest elephants that Blake studies are smaller, more elusive creatures. Only recently identified as their own species, forest elephants live in Africa's impenetrable jungle, and their behavioral patterns—even their numbers—are almost entirely unknown.

As part of a continent-wide elephant census that began with the conservation efforts, the Wildlife Conservation Society and the European Economic Community contracted to estimate the elephant population in the north of the Republic of Congo. The north was then almost entirely unexplored but had recently been carved into forest blocks designated for European logging interests. Michael Fay, an American botanist and former Peace Corps volunteer who was studying western lowland gorillas, was hired to conduct the survey. Today, Fay is known for having made a 1,200-mile "megatransect," a trek from Nouabalé-Ndoki to the coast of Gabon. But in 1989, Fay was just an adventurous graduate student and Nouabalé-Ndoki merely Brazzaville's name for an unexplored logging concession. Fay traversed Nouabalé-Ndoki with a group of Bangombe Pygmies. In the interior they found large numbers of forest elephants, western lowland gorillas, and chimpanzees that were unafraid of humans. Chimps are hunted everywhere in Africa, and their lack of fear in this instance led Fay to conclude that he and his team were the first humans they had ever seen. He decided that Nouabalé-Ndoki—unspoiled, vast, and teeming with wild animals—would make an ideal national park. Working with Amy Vedder and William Weber, directors of the Wildlife Conservation Society's Africa program, Fay wrote a proposal for a park that WCS, the World Bank, and the U.S. Agency for International Development agreed to fund. In a dramatic gesture that pleased conservationists, the government of Congo withdrew Nouabalé-Ndoki from the list of logging concessions. In December of 1993 it became a national park, with Michael Fay as its first director.

EARLY IN HIS TENURE, Fay recruited Blake to study wildlife at Nouabalé-Ndoki. In 1990, Blake had come to Brazzaville to work in an orphanage for gorillas whose parents had been killed in the bushmeat trade. In those days, Blake hung out with a group of De Beers diamond merchants. His best friend ("a cracking bloke") was an arms trader. He drank a lot of vodka, raced the orphanage car around Brazzaville, and ran a speedboat up and down the Congo River. But by 1993, Blake was ready for a change. When Fay asked him to work in the new park, Blake quickly accepted. He started as a volunteer. Fay remembers that he showed up "clad from head to toe and carrying an enormous green backpack that must have weighed five thousand pounds." In contrast, Fay had evolved a style of jungle travel that involved bringing Pygmies and packing light—one pair of shorts, Teva sandals, no shirt; he would wear the same clothes every day, wash them every night, and wrap blisters and cuts with duct tape. Blake rapidly adopted Fay's style and soon became, as Vedder puts it, Nouabalé-Ndoki's "wild-forest guy."

On his early surveys of the new park, Blake explored an elaborate network of elephant trails that crisscross the forest. Some trails were as wide as boulevards, and each seemed to have a purpose: one led to a grove of fruit trees, another to a river crossing, another to a bathing site. These trails existed only where there were no humans around to disrupt the elephants' lives. Outside the park, where there were human settlements, the trails vanished. Blake became certain that in the trail system was a map of the ecological and psychological mysteries of forest-elephant life. In 1997 he enrolled in the Ph.D. program at Edinburgh and began his thesis on the elephants of Nouabalé-Ndoki. "Elephants are kingpins of forest life," Blake says. "I have come to feel that if you could understand elephants you could really understand what was going on throughout the forest. Here's this bloody great big animal. It's disappearing, and we know bugger all about it."

In the years since he began his study, Blake's work has acquired a new sense of urgency, and this is one of the reasons he's invited me to join him on his long walk. In 1997, just as Blake was beginning his research, a civil war erupted in Brazzaville when the then president, Pascal Lissouba, sought to disarm a tribal faction from the north. Protracted firefight leveled what had been one of Central Africa's few intact cities; 10,000 to 12,000 people were killed in Brazzaville alone. The violence also spread to rural areas, where a third of the country's population was displaced and uncounted numbers were killed. Many Congolese fled their villages and hid in the forest, where they died of disease or starvation while trying to subsist off wild game.

"People did a lot of atrocious things and got away with them," Blake says. "Every Tom, Dick, and Harry had an AK-47. You'd go into a tiny village and half a dozen sixteen-year-olds would come strutting down the street with bandannas and automatic rifles." The war led to more hunting. Although the park itself was spared, largely because of its remoteness, the surrounding elephant population, as Blake puts it, "got hammered."

This history has contributed to Blake's conviction that the isolation—indeed the very existence—of places like Nouabalé-Ndoki is imperiled. As we've traveled, I've noticed a certain desperation on his part, as if he were convinced that whatever he doesn't learn about the elephants on this trip will never be learned—and that all there is to know about forest elephants will be irrevocably lost.

"FRESH DUNG!" Blake exclaims. He sheds his daypack and pulls out his waterproof notebook. With a ruler, he measures the diameter of the dung pile (which looks like an oversized stack of horse manure), cuts two sticks, and begins to separate seeds from the undigested roughage.

We're four days up a wide-open elephant trail along the Mokala River. The trail is thick with dinosaur-sized elephant prints. There are also hoof marks of red river hogs; the seldom seen giant forest hog, which grows to 600 pounds; and a pangolin, a 75-pound nocturnal consumer of ants and termites that is covered in dark-brown scales that look like the shingles on a roof; as well as leopard prints and both rear foot and knuckle prints of a big gorilla. Overhead, troops of monkeys chatter and scold: spot-nosed guenons, gray-cheeked mangabeys, and the leaf-eating colobus. In spite of all the tracks and animals we've come across, however, we've found little evidence that elephants have been here recently.

We travel each day with one of the Bayaka acting as a guide while the other Bayaka and Bantus, who tend to be boisterous on what for them is a junket into the wilderness, cavort well behind us so that they don't scare away the animals.

On this day, Blake's old friend Lamba has taken the lead, followed by Blake, and then me in the Gabon-viper slot. Lamba crouches over the dung pile while Blake isolates four types of seeds in it. Three of the four, he says, are dispersed only by elephants. One of these is the seed of a bush mango.

Lamba tells Blake that we're not seeing elephants along the trail because they've left the river for the hills, where the wild mangoes are bearing fruit.

"Most fruits are produced in fixed seasons," Blake says to me. "But there seems to be no pattern here with mangoes. They fruit whenever. It would be great if we could find lots of fruiting mangoes and lots of elephant signs. That's the kind of thing we're looking for, a few indicators of what moves elephant populations."

The most obvious explanation of what moves elephants is food, and Blake's research involves making a thorough study of the plants we encounter as well as chasing down feeding trails. We stop every twenty minutes so that he can make botanical notes. In order to create a definitive survey, Blake always follows the same route, varying it only when he makes side trips down feeding trails. He carries a Global Positioning System, a handheld device that translates satellite signals into geographic coordinates and which Blake uses to record the exact location of his observations. The Bayaka take care of navigation. Blake also carries a palm-sized computer, into which he enters his data. The use of such technology is new in wildlife biology. As Richard Ruggiero puts it, "[Blake]'s the first to use GPS and satellites to successfully look at the long-term movements of elephants in the forest. He's collected data no one else has looked at before."

But none of this matters if we don't see elephants. Despite Blake's estimate that as many as 3,000 elephants use the park, the animals themselves elude us. They're hard to see because they are agile and fast: Forest elephants grow to nine feet at the shoulder and weigh up to 8,000 pounds but move with surprising stealth, thanks to a pad of spongy material on the soles of their feet, which dampens the sound of breaking branches. The elephants also communicate by using infrasound, a frequency below the range of human hearing. Once elephants have determined that intruders are present, they can warn one another over significant distances—without humans detecting the exchange.

Blake has attempted to make elephants easier to find in a number of ways. In the fall of 1998, he received a grant from Save the Elephants, a foundation run by noted elephant conservationist Iain Douglas-Hamilton, to outfit several elephants with GPS collars. Blake and Billy Karesh, a Wildlife Conservation Society field veterinarian, went deep into the forests of Central Africa with a high-powered tranquilizing rifle; they managed to sedate two elephants near

Nouabalé-Ndoki and put collars on them. One of the collars never worked, but the second, placed on a female, worked for a month, long enough to trace the elephant's movements outside the protected forest.

The fewer signs we see of elephants, the more restless Blake becomes. "Amazing, isn't it," he muses. "Absolute bugger all."

On the fifth day, as we're walking along a ridge above the Mokala, Blake hears a branch snap. Zonmiputu is our guide. He is a quiet man, about five feet tall, an inch or two shorter than the rest of the Pygmies, and perhaps forty years old.

"*Ndzoko,*" Zonmiputu whispers. Elephant.

Quietly he puts down his pack, indicates the elephant's direction with his machete, and leads us at a crouch through the thick underbrush. After thirty-five yards, Zonmiputu stops and points out a shadowy shape looming twenty yards away. It is a young-looking bull, about eight feet at the shoulder, with deep chocolate-colored skin. I can see its brown tusks waving as it reaches up with its trunk and rips branches out of the surrounding trees. We approach. Blake hands me his binoculars. The elephant is now fifteen yards away, and I'm focused on its eye—a startling sight, sunken in the wrinkled skin, bloodshot; it seems to peer out from another epoch, as if it were looking forward at some huge, unfathomable span of time.

"A young bull," Blake whispers. "Perhaps twenty years old."

The bull senses that we're near, lifts its trunk toward us, and crashes off into the forest.

IN THE EVENING, sitting around the camp after dinner, I ask Blake to ask the Bayaka if any of them has ever killed an elephant. I know that Pygmies have traditionally hunted elephants with spears. As Blake relays the question, the Bayaka stiffen. It's illegal to kill elephants. They don't know why I'm asking, and they all say no—unconvincingly. All, that is, except Lamba. Blake refers to Lamba as "Beya," the Bayaka word for giant forest hog, because he has, as Blake puts it, "scabby habits." Having made this trip several times together, Lamba and Blake are perpetually laughing at each other, and, in front of Blake, Lamba doesn't bother to dissemble. He's killed three elephants with his spear, he tells us. He stalked the elephants and speared them in the gut. When necessary, he'd spear one a second time in the foot to prevent it from running.

For one of these elephants, which a Makao Bantu hired him to kill for its tusks, Lamba was paid an aluminum cooking pot. For another, he received a pair of shorts.

"This for a hunt that would have taken him weeks," Blake says hotly.

I ask Lamba whether he has any fear while hunting an elephant.

No, he doesn't, he responds, even though elephants can kill hunters. Gorillas, however, scare him. A mature male—a silverback—can grow to over 400 pounds. He knows three Pygmies who've been killed while stalking gorillas.

"And do the Bayaka kill *people*?" Blake asks.

This elicits nervous laughter. Cannibalism is not unknown in this region, though no one has ever accused the Bayaka of eating people. But we're not far from Bangui, where, in modern times, Emperor Bokassa is said to have served human flesh at state dinners. Blake tells me that the first Frenchman to arrive in Makao in 1908 was eaten. "We found records of it in the colonial archives in Paris," he tells me. (Later, when looking in vain for a copy of the document at park headquarters, I turn up a similar complaint from another colonist whose son had been eaten in a nearby village.)

When the laughter dies down, we hear a roar in the hills. It's a gorilla beating its chest.

TALKING TO MY FELLOW TRAVELERS requires several stages of translation. Most of our conversation is in Lingala, which Blake, the Bayaka, and the Bantus all speak. In addition to Lingala, however, the Bayaka speak Kaka, the Ubangui language of Makao Bantus, and Sango. Their own language—Bayaka—is Bantu-based, and if the Bayaka ever spoke an independent, non-Bantu language, it disappeared after the Bantu migration into the region thousands of years ago. As we progress farther into the forest, the Pygmies use words for plants and animals that are so specific they may be relics of an older Bayaka, the ancestral language of a forest-based people.

"There are 4,000 to 5,000 plants in this forest," Blake says one day. "I know the botanical names of perhaps 400. Mossimbo knows the Bayaka names for probably twice that. Zonmiputu knows even more."

What the language gap means is that if I want to ask the Bayaka a question, I have to first ask Blake in English, who then translates it into Lingala, which often sets off a discussion in Bayaka, which is summarized in Lingala to Blake, who finally gives it back to me in English. Meaning is distorted—lost—in the process. My frustration rises as I gradually realize that not only do the Bayaka speak several human languages but they can also summon wild animals.

We are walking under a troop of monkeys one day when Lamba begins to whistle, a loud, repeated screech in imitation of an African crowned eagle, a canopy predator. The monkeys are already screaming at us, but Lamba's sound throws them into a state of agitation and draws them down to the trees' lower branches. Soon the forest resounds with the thrashing of limbs and the cracking of branches, as well as grunts, whistles, and alarmed chattering, as the monkeys react to being caught between the imagined eagle above them and the indefinable hominids below.

On another occasion, Lamba crouches down and makes a nasal call that imitates the distress call of a duiker, a type of forest antelope that has adapted to the lack of browse on the tropical forest floor by eating fruit, flowers, and leaves dislodged by canopy monkeys and birds. Immediately a blue duiker— only a foot tall, one of the smallest of the forest antelopes—charges out of the undergrowth. It has big eyes and a small, round nose. When it spots us, it pulls up short, then turns around and bolts. But it can't resist Lamba's call. It returns, stops, bolts again, and comes back—until Lamba finally breaks the spell by laughing at the antelope's confusion.

Later, we come across a herd of fifteen red river hogs rooting and grunting around the forest floor. These hogs grow to 250 pounds and have small, razor-sharp tusks. Our presence makes them skittish, but they don't flee. They may never have seen men before. We stalk until the closest hog is five yards away, just over the trunk of a fallen tree. Mossimbo then begins a wheezing-pig call. The pigs freeze, dash away, and then, spellbound, return nervously, almost compulsively. Mossimbo keeps calling until he has the biggest boars lined up across the trunk from us. Staring, entranced, their faces look extraterrestrial— tufted ears, long snouts, big sensitive eyes ringed with white; they seem unable to fathom just what they're looking at. Mossimbo squeals—an alarm. The pigs' eyes bug out, and they race off into the forest as Mossimbo erupts in laughter.

To me these episodes are fragmentary glimpses of a world in which humans and animals share a symbolic language. The Bayaka take great pleasure in their mastery over the animal world, and nearly every episode of their summoning animals ends in guffaws. It's not benevolent laughter. If Blake and I hadn't been present, each of these animals would certainly have wound up in a Bayaka cooking pot—and there's something about this nasty, exhilarating confidence that is quintessentially human.

WE'VE COME to a point where we must ford the crocodile-rich Mokala. The current is swift, the river bottom sandy, and the water up to our chins. We hold our bags above the water level and, shortly after we reach the far shore, wade across a tributary and enter the park. As we climb up the bank, we enter an area of closed canopy forest where the understory is more passable and the butts of the trees are eight and nine feet in diameter, with straight boles that explode into kingdoms of filigree high above. Zonmiputu is again in the lead when he stops stock-still, turns back to us, and whispers, *"Koi."* Leopard.

Through a gap in the underbrush, we make out a pattern of dark rosettes on a brown background. The impression gradually resolves into the abdomen and haunch of a large leopard. As we watch, it glides out of the frame, its snaky tail trailing behind.

Zonmiputu crouches, clears his throat, and makes a duiker call to try to fool the leopard into coming to investigate. Through another gap, I see the leopard hesitate, then break into a run. It's gone.

Although leopards are not commonly believed to attack humans, the Pygmies claim they do. Several days earlier, Mossimbo had pointed out a pile of leopard scat filled with reddish-brown hair. Blake poked around in it long enough to discover a strange brown cylinder the size and color of a cigar butt. Using a stick, we rolled it over. I leapt back in horror. It was the top half of a finger, the nail still intact.

"Chimpanzee," Blake said.

THE EERIE DISCOMFORT of the forest is beginning to overwhelm me. One night, I'm inside my tent in the grips of a dream. I'm being suffocated by vines, buried until only my face is exposed. Slowly I'm being pulled into the earth. I awake with a start, pull out my flashlight, and check my watch. It's four-thirty in the morning. The air inside the tent is thick and stifling. Outside, water drips from leaves, unseen creatures scurry, branches snap, beasts hoot and squeak. Overhead, I feel the claustrophobic weight of tropical foliage. Tonight's dream is one of a series that has become vivid—houses I used to live in, offices I've worked in, visits with friends—and I wonder if this is what it's like to be dead. My restless spirit is haunting the places I loved.

Dawn is filtering down to the forest floor. I hear the rest of the camp stirring: the Pygmies whack their machetes into dead branches and clang our battered, soot-covered aluminum pots over the fire. I hear Blake yawning in his tent. He calls out to ask how I've slept. I lie and tell him I've slept well. But now, at six in the morning, this trip has become oppressive. Breakfast arrives: a mound of glutinous white rice covered with Moroccan sardines and the leftovers of last night's smoked-fish stew. Gloomily, I tuck in. Blake asks if I find the forest claustrophobic. I lie again and tell him no, but it's a bad line of thinking, because today in fact I do. Neither am I heartened by the fact that today we're not moving camp. While Blake goes off to do some elephant-feeding studies, I'll have to spend the day alone with the Bayaka.

The Bayaka and I leave camp around eight, cross a stream, head up into the hills, and wander, foraging for wild mushrooms, yams, seasonal fruit, a bark that tastes like garlic, a bark that serves as an antibiotic, a bark containing quinine, an edible vine in the legume family, a sapling that is said to act like Viagra—in short, whatever the forest will provide. With Blake, we follow elephant trails and walk purposefully in single file. With the Bayaka we maintain no consistent direction. My compass becomes useless. I cling to my guides.

My first Bayaka encounter occurred as Blake and I stepped from our pirogue into the waiting crowd at the riverside. A kindly-looking old Bayaka in a torn shirt stepped out from the back of the crowd and headed straight for me. He grasped my hand, stared curiously into my eyes, and wouldn't let go.

"He just wanted to see what kind of a person you are," Blake explained, once I'd pried my hand free.

It was almost as if the old Bayaka recognized me. If he had, it wouldn't have been entirely far-fetched. Douglas Wallace, the geneticist who has made a career reconstructing human migrations around the globe through rates of change in mitochondrial DNA, believes that the Bayaka are descended from a small group of Paleolithic people who once roamed across eastern Africa. Wallace and several others argue that a population genetically very close to the present-day Bayaka were the first modern humans to leave Africa some 50,000 years ago. "We are looking at the beginning of what we would call Homo sapiens," he wrote recently.

In other words, I live in New York, but I'm also the long-lost cousin of the Bayaka, the depigmented descendant of their ancestors who hiked over the horizon and never came back—until now.

Strolling through the forest, I've noticed that my cousins appear to be in a perpetual Wordsworthian idyll; they often gaze dreamily upward, as if contemplating the god that has provided them with such sylvan abundance. At one point, I convey this impression to Blake. He corrects me. "What they're looking for is not divinity but wild honey. Although, for them, it's pretty much the same thing."

Sure enough, my day with the Bayaka devolves into a honey hunt. There are several kinds of wild honey in the forest; one belongs to a stinging bee. Mossimbo doesn't take long to spot what he takes for a stinging-bee hive high overhead, sixty feet up, in a hole in a tree branch. The Bayaka rapidly build a fire and extinguish it. Mossimbo wraps the coals in a bundle of leaves, straps the bundle on his back, grabs a machete, and effortlessly shinnies up a liana along the branchless tree trunk. Soon he's vanished into the foliage, and all we can hear is his machete hacking into a tree branch. Finally he descends with two dry honeycombs.

"*Chef,*" he says, drawing on his minimal French for the first time. "*C'est fini.*"

The hive has been abandoned.

After several more hours of wandering, the Pygmies spot a more accessible stingless sweat-bee hive, climb the tree, and soon revel in honey that tastes watery, smoky. I sample it, but to me it's an off-putting soup of bark, twigs, grubs, and dead and dying bees. I leave it to the Pygmies.

After we leave the hive, the sweat bees pursue us vengefully. We're squatting down in front of a pile of bush mangoes, shucking the seeds out of the hardened pits, when I'm suddenly overcome with helplessness. A large part of my frustration comes from the language. Blake is not here to translate my questions. But I'm not just deprived of speech today; I'm also faced with the fact that the forest, which is such a source of bounty to the Bayaka, is, to me, an undifferentiated mass. I don't have the vocabulary to break this environment down into parts. There's nothing I can parse, nothing I can usefully understand. I'm completely at a loss without words. The Pygmies see that I'm wilting.

"Papa," Mossimbo says, affectionately, handing me a mango seed.

By the time we get back to camp, it's thick with tsetse and filaria flies. Tsetse flies carry sleeping sickness. Filaria flies can deposit the larva of parasitic worms in a human's bloodstream. (Blake later comes down with fly-borne elephantiasis.) One of the Bantus slaps a tsetse that is feasting on my back, leaving the dead insect lying in a pool of my blood. I think of what Blake told me when I'd been bitten earlier by a tsetse. "No one can tell me those flies can't transmit AIDS. All you need is a few viral cells. We're in the Congo, after all, and the AIDS problem is huge."

As the afternoon ends, I'm not fit for anything but crawling inside my tent.

The flies disappear at sundown, and I re-emerge for dinner. Smoked fish again. This time it's served with manioc, a cloying flour made from the tuberous root of the cassava plant. After our meal, the Bayaka pull out *djamba*—marijuana—a substance that the Bayaka value only slightly less than wild honey. As they have on many nights, the Bayaka roll the marijuana in forest leaves and inhale deeply. Tonight the ensuing hilarity seems greater than usual.

Since all the jokes are in Bayaka or Lingala, I ask Blake for explanations. The Pygmies have asked him if I'm rich, he says. Obligingly, he has told them that I am the richest man in the world.

"You're their new culture hero," he says.

I try to imagine what might have led the Pygmies to speculate about my wealth, and remember that, in addition to the Tevas I wear most days (we're constantly in and out of water), I have two pairs of sneakers, one of which I haven't even worn. Three pairs of shoes! Extravagant, prodigal—*rich.*

I retire to my tent and, crawling in, notice that the ground under the tent floor is blotched with patches of light. I have smoked the *djamba,* and the tent floor looks like a city at night seen from an airplane. Until I figure out that it's a phosphorescent fungus, this vision offers consolation, if only because it reminds me that there is a city out there, somewhere in the world. I fall asleep and have another strange Ndoki dream. A woman appears and teaches me the supernatural art of being in two places at once.

WE'VE REACHED the line of *bais* stretching from north to south that defines the center of the park's elephant life. The word *"bai"* is derived from the French *"baie,"* but it has escaped into local usage to describe a miniature savanna maintained by forest elephants in the middle of the forest. "If elephants are lost from an area," Blake says, "*bais* quickly grow over."

One afternoon, Blake and I follow the elephant trails to the Bonye River *bai.* The *bai* is big, the size of three football fields, and it's the first open terrain we've seen since leaving Makao. The afternoon light is soft and golden, playing on the riffling surface of the river as it winds through the clearing. In the water, about seventy-five yards away, are nine forest elephants—four adults and five young, three of them infants. As we watch, an old matriarch ambles out of the forest, followed by two more young and another adult female. The matriarch reaches the riverbed, kneels, drills her trunk into the white sand, and gurgles as she sucks mineral-rich water out of the streambed. When she pulls her trunk up, she sprays river water into her mouth. Upstream, a wading bird picks at the riverbed while three red river hogs browse on the marsh grass, trying to avoid the playful charges of one of the baby elephants. On the far margins a sitatunga, with its distinctive wide, splayed feet, feeds quietly.

During the next hour and a half there is a constant coming and going until we've seen thirty elephants in all. The young ones prance around and engage in mock fights, and the adults spray themselves and their children, as the sunlight flashes in the water droplets. Blake looks blissful. He creeps forward to the edge of the *bai*, quietly sets up a video camera, pulls out his notebook, and begins sketching what are the most distinctive and identifying features of individual elephants: their ears. He'll exchange these later with Andrea Turkalo, a forest-elephant researcher who is studying the social structures of elephant herds in a Dzanga Sangha *bai* across the border in the Central African Republic.

I sit on a fallen tree trunk, relieved, enjoying the light. Elephants are members of the ancient, highly successful order of Proboscidea, which, historically, has contained almost 200 different trunked and tusked species, including mastodons and mammoths. Beginning 50 million years ago, and as recently as the late Pleistocene, 10,000 or so years ago, proboscideans roamed the globe. Mastodons and mammoths grew up to fifteen feet. But there were also pygmy elephants. (A four-foot-tall elephant, *Elephas falconeri*, survived on the Greek island of Tilos until a little over 4,000 years ago. A dwarf mammoth lived on Wrangel Island off Siberia until 1700 B.C.) Then, toward the end of the Ice Age, elephants died off en masse, and today only two species survive—the Asian elephant, *Elephas maximus*, and its bigger cousin, the African elephant, *Loxodonta africana*. Both of these species evolved in Africa, but *Elephas* moved into Asia and then became extinct in its home range. Some argue that only then—about 40,000 years ago—did *Loxodonta africana*, which had been exclusively a forest creature, emerge to seize the open savanna.

The mass proboscidean die-off was part of the mysterious and more general Pleistocene extinctions. Sometime between 10,000 and 25,000 years ago, all mammals weighing more than a ton—as well as many lighter than that—disappeared from Europe, Asia, and the Americas. This is an old story. But the other story is that some elephants survived—as a miracle, emissaries from the prehistoric world.

FROM BONYE BAI we head south to Little Bonye *bai*, Mabale *bai*, and, ultimately, Mingingi *bai*, the epicenter of elephant life in the park. Blake points out the various fruit trees associated with elephant trails. The most conspicuous of these, he says, is *Duboscia macrocarpa*, a large tree with an almost gothically fluted trunk. Virtually every duboscia we see stands at the intersection of several elephant trails, gracefully alone in a clearing made by fruiting-season elephant traffic to the tree. Another regular tree along the trails is *Omphalocarpum elatum*, which has fruit growing out of the side of its trunk. The fruit is encased in a heavy, hard-shelled ball—the size of a medicine ball—which the elephants like well enough to dislodge by ramming the tree with their heads.

"The importance of fruit trees for forest elephants has only recently been acknowledged," Blake says. "And that's because almost all elephant research has been based on savanna elephants, which eat very little fruit. In fact, many conclusions drawn from savanna-elephant research are simply not applicable to forest elephants. It's always amazing to me that elephants get lumped in categories the way they do."

As we walk, Blake confesses to me his obsession with the rocker Chrissie Hynde, and in particular with her song "Tattooed Love Boys." And one evening, after we arrive in camp, Blake spots Lamba sprawled across his bags. Blake takes his daypack, lifts it over his head, stands over Lamba as if to hurl it down, and recites:

> Run to the bedroom.
> In the suitcase on the left,
> You'll find my favorite axe.
> Don't look so frightened.
> This is just a passing phase,
> One of my bad days.
> Would you like to watch TV?
> Or get between the sheets?

Lamba is baffled but, with the rest of the Bayaka, laughs nervously. " 'One of My Turns,' by Pink Floyd," Blake explains to me. "You never heard *The Wall* concert they played in Berlin, did you? There's that whole debate about stadium concerts. I'm not that big a fan of stadium concerts, but that was a great concert."

As we're ducking under some vines, we see our first snake. It's in the branches overhead—a big, evil-looking thing nearly five feet long. Blake can't identify it, but it's not one of the famously poisonous snakes of this region—not a boomslang, not a black mamba, not one of the several cobras. The Bayaka give us their name for it, say it's bad, and seem anxious to get away from it. "Can you imagine how many others we haven't seen?" says Blake.

Shortly afterward we scare a leopard off a fresh-killed duiker. The duiker's entrails are ripped out, but it's still warm. The Bayaka tie up the duiker and take it along for our dinner.

We're wading in the sandy shallows of the Mabale River when we discover a dead baby elephant. It's a gruesome and disturbing sight; the elephant, the size of a pony but stouter, is half submerged, covered in flies, and leaking blood from its trunk.

At this same spot last evening, we saw one of the elephants Blake had collared two years earlier now standing in the river, still wearing her nonfunctioning transmitter. Considering the size of the park, this was quite a coincidence. And not only did we see her yesterday but we saw her with a young elephant following close behind her; at the time she was collared, she had been pregnant, and this young elephant looked about a year old—the appropriate age. Now, it seems, that baby is dead.

"Hell of a thing," says Blake, pacing back and forth. "Hell of a thing." He picks up the baby elephant's trunk, lets it flop back into the water, and examines the tiny tusks and the toenails on each foot. He picks up a stiff leg and turns the little creature over, looking for some telltale sign of what killed it, but he can't find anything except a group of puncture wounds on the animal's chest. With snakes on my mind, I suggest that the wounds might be the result of Gabon-viper bites—and that the elephant may have died of hemolytic bleeding. Blake is unimpressed but seems distressed that he can't come up with an explanation. He frets, hovers, pulls out his notebook and takes notes, gets his video camera and shoots pictures. He is reluctant to leave.

Looking over the little creature, dead of unknown causes, I'm struck again with a sense of being dead—the idea that this lifeless body could be mine. After a quarter of an hour, I persuade Blake to give up his forensics, and we start up a trail—only to turn back ten minutes later. He has decided the puncture wounds are the result of a leopard attack.

"If we could demonstrate that a leopard could kill a baby elephant, it would be quite a thing," he says.

We wade back into the river. Blake pulls out his knife and makes precise incisions along the puncture wounds, two of which go straight through the elephant's chest and into its lungs. A punctured lung could be the source of the bleeding through the trunk. The elephant could have drowned in its own blood. Blake's hypothesis about the leopard suddenly seems plausible.

"Hell of a thing. Hell of a thing," Blake repeats to himself, still agitated, but in much better spirits now that he's arrived at a theory. It occurs to me that science is formidable, and not merely for its accomplishments but because faith in reason leads people to brave treacherous environments like this one.

WE CAMP ALONG the Mingingi River, a mile or so below the *bai*. The day is sultry, buggy. Thunderclaps rumble across the distant forest, and late in the afternoon we're drenched by a brief downpour. But the weather clears overnight, and I awake at three in the morning to gorilla calls echoing up the valley, elephants trumpeting from the *bai* above, and moonlight illuminating the side of my tent.

In the morning we head up to the *bai*. The approach paths are wide and parklike, and the landscape has been designed by elephants. They have dug bathing pools out of the hillsides. The underbrush has been cleared of patches of forest, and tree trunks are swollen to exotic shapes from elephants having picked away their bark. We find a meadow surrounding a highly polished termite nest—an elephant rubbing post surrounded by the marks of heavy traffic and worn down so far that it looks somehow like a public monument.

We creep forward toward the edge of the *bai*, a huge open space of marshy grassland and isolated clumps of trees. A shower has just passed and the mist is lifting off the forest all around. Swallows are dipping in the river. A white palm-nut vulture with its hooked yellow beak is perched on a dead tree limb. A single bull is drinking from a pool.

Just as we're preparing to walk out into the river, nine bongos—large forest antelopes—emerge out of the underbrush and wade into the middle of the *bai,* tails flicking, sides adorned with vertical white stripes, their celestial-looking horns curving gracefully skyward.

LAMBA HEARS what he says is a yellow-backed duiker, the largest of the duikers. We're beyond Mingingi, in the center of the park. We squat while Lamba calls. No response. He calls again. A stick snaps. Silence. Lamba calls a third time. Another stick snaps, off to our right. I wheel around and see two heads duck quickly behind a termite mound. It's a strange, stealthy gesture. The heads are humanlike. We're not the only ones stalking; we're being stalked. Chimpanzees, thinking they're going to find a wounded duiker, have instead found their nearest primate cousins.

"It's just the lads," Blake says, "checking us out."

Along with baboons—and of course humans—chimpanzees are the only primates who regularly kill other mammals. In Nouabalé-Ndoki, chimpanzees set methodical ambushes for the leaf-eating colobus monkeys and even for duikers. They also scavenge other meat—including pigs—and Blake tells me that in the past he has called in chimpanzees by hiding behind tree roots and making duiker calls.

"When they respond to the duiker call, they come for the kill," Blake says. "The males are quite a sight with their tails up and their hair standing on end.

They come whipping around the tree root, see us, and just deflate. They've never seen humans before. I've had one sit and stare at me for five minutes."

In his book *The Third Chimpanzee*, the physiologist Jared Diamond argues that humans are close enough to chimpanzees to properly be thought of as a third chimpanzee species (after chimpanzees and bonobos). The DNA of chimpanzees is 98.4 percent the same as ours, and of the remaining 1.6 percent, most is insignificant. The meaningful genetic differences could be focused in as little as one tenth of one percent; they account for the genes that lengthened our limbs (allowing us to walk upright and use tools) and, more importantly, altered, as Diamond puts it, "the structure of the larynx, tongue, and associated muscles that give us fine control over spoken sounds." Indeed, a group of scientists have recently isolated a single gene that may underlie the human ability to speak. These scientists are presently trying to determine when this gene evolved. One theory dates it to only 50,000 years ago—around the time the ancestral Bayaka left Africa and set out to explore the world.

We're in chimpanzee territory now, and after our stalking encounter we find signs of chimps everywhere. We hear them pounding on tree trunks and howling like coyotes out in the forest. We see their skillfully made nests in the trees and the ingenious traps they've set at termite nests, but we don't see the chimps themselves. Noticing that I've become preoccupied with spotting a chimp, Lamba volunteers that the Bayaka make a chimp-hunting charm, but when I ask him about it he averts his eyes. The next day one of the Bantus speaks up. "*Chef,*" he says to Blake, "Lamba was lying. There's no chimp-hunting charm—only a gorilla-hunting charm."

A few nights later, Zonmiputu strips a liana down into fine strands and dries the strands over the fire. ("It's *Manneophyton fulvum,* the liana they use for making hunting nets," Blake explains.) Zonmiputu tosses the mass of shredded vine to Manguso, another of the Bayaka. Taking the mass of vine with him, just at sunset, Manguso climbs into the lower branches of a tree. He makes gentle sounds, gorilla sounds—imploring noises, soft exclamations, sounds of surprise—all the while weaving whatever he's expressing into a rope.

"You do it this way," Zonmiputu explains, "so you can get the gorilla up in a tree."

It's the gorilla-hunting charm.

"Not every Bayaka knows how to make this," Blake says to me. "These people are disappearing as fast as the elephants, and their knowledge is disappearing with them."

Two days later, I'm wearing the charm bandolier-style across my chest when Zonmiputu sees me. He looks alarmed. He's made the gorilla charm for me, but one of the other Bayaka is supposed to wear it. Such things are supposed to be worn only by initiated Bayaka—but we can't, of course, talk to each other, and I only learn about this prohibition later. Zonmiputu sends one of the Bantus to explain that if I come across an elephant, I must take it off. Otherwise the elephant will become mean.

We smell the gorillas before we see them. There's a dusky odor along the trail. The gorillas are just ahead of us, and apparently they smell us. There's a loud crash of tree branches, and a silverback barks, then ignominiously flees. A female with an infant on her back and two juveniles are caught in the trees. For the next ten minutes, they try to muster the nerve to descend and flee. Eventually, the mother, the infant, and one of the juveniles make death-defying leaps to the ground and run off into the underbrush. The remaining juvenile stays behind, defiantly pounding his little chest until we move along and leave it in peace.

"They're in for a shock when the loggers get here," Blake says.

WE'RE NOW OUT of the park and in the Pokola logging concession, which is leased to the German-owned, French-managed company Congolaise Industrielle des Bois (CIB). Blake, who was jubilant while in the forest, now seems depressed.

"Our wilderness walk is over as far as I'm concerned," he says. "We're now in the realm of man."

The prospecting line, however, is only the first sign of what Blake refers to as the park's "biggest land management issue"—industrial logging. CIB now has the rights to two of the three concessions surrounding the park, and over the next twenty years the entire forest surrounding the park will be selectively logged. What this means, Blake says, is that in twenty years the only intact forest in the north of Congo will be Nouabalé-Ndoki.

Logging itself is not the most dangerous threat to wildlife. Loggers in the region generally confine themselves to removing only two species of African mahogany that bring high enough prices on the European market to justify the expense of transporting them. (A single African mahogany log might bring $4,000 on the dock at a European port.) The additional light brought to the forest floor as trees come down may even promote the growth of ground ferns favored by many large mammals, including elephants, and logging may *increase* densities of certain animals. But by building roads, bringing in thousands of foreign workers, and creating a cash economy, logging has invariably led to uncontrolled killing of animals—poaching. CIB is working with the Wildlife Conservation Society and the Congolese government to develop wildlife management within logging concessions and to control poaching, but it remains an ominous situation.

"The big issue," Blake told me, "is for the logging companies to take responsibility for hunting in their concessions. It's not a feasible argument for us to say they shouldn't be here—Congo needs revenue, and we'd be laughed out of the country. Controls on hunting, the prohibition on the export of bushmeat, and the importation of beef or some other source of protein are about the extent of our demands on the company."

WE DESCEND the Bodingo peninsula, an elevated ridge of land south of the park's border that runs down into the Likouala aux Herbes swamps. We soon discover that more than a prospecting line has been cut through the forest. CIB has surveyed much of the peninsula, marking off the commercially valuable trees with stakes in the ground. The prospectors appear to have been accompanied by a party of Pygmy hunters. We see abandoned snares and places where trapped animals have struggled to free themselves by digging holes in the ground and raking trees with their claws in attempts to escape.

The forest has been cut up in a grid, letting in light, leaving it curiously thin. Taking in the devastation around us, I realize that what's lost when a forest is cut is the weight of evolutionary history, the whole sequence of life, all the voices that the Bayaka can still understand—the voices that existed in nature before we other primates found a way to describe, and circumscribe, the world around us.

Blake is studying an African mahogany that's been marked for harvest. Its dense trunk, which is ten feet in diameter and has oaklike bark, soars upward toward the canopy. "That tree may be 900 years old," he says. "Soon it will be gone. Just like that."

We continue down the peninsula and launch off into the swamps. Tsetse flies are in evidence, along with sword grass, thorn forest, army ants. We sprint through the ant columns. We sleep on patches of raised earth, bathe in mud puddles, and drink coffee-colored water out of stagnant pools. One day Lamba finds a greenish-water-filled excavation—the home, he explains, of an African dwarf crocodile. He squats down and makes a birdlike sound. Soon eight little crocodile heads nervously broach the algae-green surface, their elevated eyes popping up like bubbles.

I am walking behind Zonmiputu when I look up and spy a spider the size of a dessert plate crawling up his back toward his hair and his collar. "Putus!" I shout, using his nickname. Zonmiputu freezes. This is distressing. Zonmiputu is supposed to be invulnerable. I run up behind him, intending to brush the spider off. But the spider has a furrowed, lethal-looking body and strong hairy legs that are tensing as if it is preparing to leap. I grab a stick and whisk the spider into the bushes. Zonmiputu turns around, looks at my spider pantomime, grimaces, shudders, and hurries back along the trail.

In Nouabalé-Ndoki there is always the unnerving sensation that something is watching you. A mongoose creeps through the underbrush; a tree snake twirls along a branch. Today, as we scramble over root snarls, plunge thigh-deep through pools of mud, and approach Terre de Kabounga, the end of our walk, we come across the fresh trail of a crocodile and then hit something that really stops us: a human footprint in the mud. It's so fresh that it's still filling with water. Someone has spotted us, and he's hiding.

The Bayaka find the trail and follow it. We hit dry land and soon hear a woman singing, a meandering, flutelike voice. A tall, graceful Bantu woman, clad in brightly colored wax-print African fabrics, her hair in cornrows, is gathering firewood and, though she has seen us, defiantly continues her song. Before long a husband emerges. He's a square-shouldered, handsome man, the schoolteacher, he tells us, from the nearby town of Bene. He hasn't been paid in three years, so he closed down the school and left his students, the future of Africa, to fend for themselves. He moved out into the swamps, along with a good part of the rest of this region's shattered population, to smoke fish and hunt bushmeat.

We follow the schoolteacher and his wife to their camp. It's filled with fish and hung with shotguns. Other relatives come out of the forest to stare at us in wonder. They direct us to a path that leads, an hour later, to the cut-over edge of the forest. We emerge onto a red-clay road, blinking and squinting in the harsh, flat light of the open road. The heat, unfiltered by the forest, hits like a blast furnace. We shake hands in a gesture of shared congratulation, but the triumph feels hollow. We've been dreaming of the human world, but now that we've arrived it's disorienting.

We walk for hours. Late that afternoon a big, flatbed Mercedes drives up. The driver is so drunk he can barely stand. The ten of us find space in back among twenty-seven other passengers, sacks of manioc tubers, baskets of smoked fish, mounds of edible leaves, and the carcasses of several dead duikers. Soon we're being carried off toward the logging town of Pokola at such high speeds that at times the big truck seems to go airborne. I offer a silent prayer that, having survived a month in the forest, I won't be killed in a car crash.

A decade ago, Pokola was a tiny fishing village on the Sangha River. It's now a sprawling shantytown built of scrap mahogany. In its busy market, the Bayaka spend their pay outfitting themselves in bright sports clothes until they look, in Blake's words, "like Cameroonian soccer stars." Blake and I drink wine with the French logging managers inside their fenced-off compound. I pull the tick out of my nose. Before I know it, I'm back in New York, where I am treated for schistosomiasis, amoebic dysentery, and whipworms.

BLAKE'S NINTH AND last "long walk" capped the first phase of his doctoral research and gave him the data to begin writing his thesis. Since our trip he has returned to the interior of Nouabalé-Ndoki several times to collar more elephants and collect data to support his argument that by disseminating the seeds of forest-fruit trees, elephants play a crucial role in the evolution of Central African forests.

But in the interim, civil war has broken out again, and Blake reports that since our trip all of the remaining concessions in the north of Congo have been leased to logging companies. A new sawmill is being built north of the park, and a logging road now runs straight into Makao. Another road cuts across the Bodingo peninsula close to the park's southern border. The place where we saw the gorillas, Blake reports, is already a lacework of logging trails. "The civil war was a disaster for the country," he says. "If there'd never been a civil war, the government might have been more open to conservation. Now development and reconstruction have become the country's highest priority.

"In many ways," he says, "what we saw is already gone."

CHARLES C. MANN

1491

FROM *THE ATLANTIC MONTHLY*

Recent research into the natural history of the pre-Columbian Americas is turning up some provocative findings about the size of the Indian population and the sophistication of its culture. A group of archaeologists and anthropologists is threatening to overturn many cherished ideas about the Indians' civilization—and creating controversy in the process, as Charles C. Mann finds out.

The plane took off in weather that was surprisingly cool for north-central Bolivia and flew east, toward the Brazilian border. In a few minutes the roads and houses disappeared, and the only evidence of human settlement was the cattle scattered over the savannah like jimmies on ice cream. Then they, too, disappeared. By that time the archaeologists had their cameras out and were clicking away in delight.

Below us was the Beni, a Bolivian province about the size of Illinois and Indiana put together, and nearly as flat. For almost half the year rain and snowmelt from the mountains to the south and west cover the land with an irregular, slowly moving skin of water that eventually ends up in the province's northern rivers, which are sub-subtributaries of the Amazon. The rest of the year the water dries up and the bright-green vastness turns into something that resembles a desert. This peculiar, remote, watery plain was what had drawn the researchers' attention, and not just because it was one of the few places on earth inhabited by people who might never have seen Westerners with cameras.

Clark Erickson and William Balée, the archaeologists, sat up front. Erickson is based at the University of Pennsylvania; he works in concert with a Bolivian archaeologist, whose seat in the plane I usurped that day. Balée is at Tulane University, in New Orleans. He is actually an anthropologist, but as native peoples have vanished, the distinction between anthropologists and archaeologists has blurred. The two men differ in build, temperament, and scholarly proclivity, but they pressed their faces to the windows with identical enthusiasm.

Dappled across the grasslands below was an archipelago of forest islands, many of them startlingly round and hundreds of acres across. Each island rose ten or thirty or sixty feet above the floodplain, allowing trees to grow that would otherwise never survive the water. The forests were linked by raised berms, as straight as a rifle shot and up to three miles long. It is Erickson's belief that this entire landscape—30,000 square miles of forest mounds surrounded by raised fields and linked by causeways—was constructed by a complex, populous society more than 2,000 years ago. Balée, newer to the Beni, leaned toward this view but was not yet ready to commit himself.

Erickson and Balée belong to a cohort of scholars that has radically challenged conventional notions of what the Western Hemisphere was like before Columbus. When I went to high school, in the 1970s, I was taught that Indians came to the Americas across the Bering Strait about 12,000 years ago, that they lived for the most part in small, isolated groups, and that they had so little impact on their environment that even after millennia of habitation it remained mostly wilderness. My son picked up the same ideas at his schools. One way to summarize the views of people like Erickson and Balée would be to say that in their opinion this picture of Indian life is wrong in almost every aspect. Indians were here far longer than previously thought, these researchers believe, and in much greater numbers. And they were so successful at imposing their will on the landscape that in 1492 Columbus set foot in a hemisphere thoroughly dominated by humankind.

Given the charged relations between white societies and native peoples, inquiry into Indian culture and history is inevitably contentious. But the recent scholarship is especially controversial. To begin with, some researchers—many but not all from an older generation—deride the new theories as fantasies arising from an almost willful misinterpretation of data and a perverse kind of political correctness. "I have seen no evidence that large numbers of people ever lived in the Beni," says Betty J. Meggers, of the Smithsonian Institution. "Claiming otherwise is just wishful thinking." Similar criticisms apply to many of the new scholarly claims about Indians, according to Dean R. Snow, an an-thropologist at Pennsylvania State University. The problem is that "you can make the meager evidence from the ethnohistorical record tell you anything you want," he says. "It's really easy to kid yourself."

More important are the implications of the new theories for today's ecolog-ical battles. Much of the environmental movement is animated, consciously or not, by what William Denevan, a geographer at the University of Wisconsin, calls, polemically, "the pristine myth"—the belief that the Americas in 1491 were an almost unmarked, even Edenic land, "untrammeled by man," in the words of the Wilderness Act of 1964, one of the nation's first and most impor-tant environmental laws. As the University of Wisconsin historian William Cronon has written, restoring this long-ago, putatively natural state is, in the view of environmentalists, a task that society is morally bound to undertake. Yet if the new view is correct and the work of humankind was pervasive, where does that leave efforts to restore nature?

The Beni is a case in point. In addition to building up the Beni mounds for houses and gardens, Erickson says, the Indians trapped fish in the seasonally flooded grassland. Indeed, he says, they fashioned dense zigzagging networks of earthen fish weirs between the causeways. To keep the habitat clear of un-wanted trees and undergrowth, they regularly set huge areas on fire. Over the centuries the burning created an intricate ecosystem of fire-adapted plant species dependent on native pyrophilia. The current inhabitants of the Beni still burn, although now it is to maintain the savannah for cattle. When we flew over the area, the dry season had just begun, but mile-long lines of flame were already on the march. In the charred areas behind the fires were the blackened spikes of trees—many of them, one assumes, of the varieties that activists fight to save in other parts of Amazonia.

After we landed, I asked Balée, Should we let people keep burning the Beni? Or should we let the trees invade and create a verdant tropical forest in the grasslands, even if one had not existed here for millennia?

Balée laughed. "You're trying to trap me, aren't you?" he said.

Like a Club Between the Eyes

ACCORDING TO FAMILY LORE, my great-grandmother's great-grandmother's great-grandfather was the first white person hanged in Amer-ica. His name was John Billington. He came on the *Mayflower*, which anchored off the coast of Massachusetts on November 9, 1620. Billington was not a Puri-tan; within six months of arrival he also became the first white person in

America to be tried for complaining about the police. "He is a knave," William Bradford, the colony's governor, wrote of Billington, "and so will live and die." What one historian called Billington's "troublesome career" ended in 1630, when he was hanged for murder. My family has always said that he was framed—but we *would* say that, wouldn't we?

A few years ago it occurred to me that my ancestor and everyone else in the colony had voluntarily enlisted in a venture that brought them to New England without food or shelter six weeks before winter. Half the 102 people on the *Mayflower* made it through to spring, which to me was amazing. How, I wondered, did they survive?

In his history of Plymouth Colony, Bradford provided the answer: by robbing Indian houses and graves. The *Mayflower* first hove to at Cape Cod. An armed company staggered out. Eventually it found a recently deserted Indian settlement. The newcomers—hungry, cold, sick—dug up graves and ransacked houses, looking for underground stashes of corn. "And sure it was God's good providence that we found this corn," Bradford wrote, "for else we know not how we should have done." (He felt uneasy about the thievery, though.) When the colonists came to Plymouth, a month later, they set up shop in another deserted Indian village. All through the coastal forest the Indians had "died on heapes, as they lay in their houses," the English trader Thomas Morton noted. "And the bones and skulls upon the severall places of their habitations made such a spectacle" that to Morton the Massachusetts woods seemed to be "a new found Golgotha"—the hill of executions in Roman Jerusalem.

To the Pilgrims' astonishment, one of the corpses they exhumed on Cape Cod had blond hair. A French ship had been wrecked there several years earlier. The Patuxet Indians imprisoned a few survivors. One of them supposedly learned enough of the local language to inform his captors that God would destroy them for their misdeeds. The Patuxet scoffed at the threat. But the Europeans carried a disease, and they bequeathed it to their jailers. The epidemic (probably of viral hepatitis, according to a study by Arthur E. Spiess, an archaeologist at the Maine Historic Preservation Commission, and Bruce D. Spiess, the director of clinical research at the Medical College of Virginia) took years to exhaust itself and may have killed 90 percent of the people in coastal New England. It made a huge difference to American history. "The good hand of God favored our beginnings," Bradford mused, by "sweeping away great multitudes of the natives . . . that he might make room for us."

By the time my ancestor set sail on the *Mayflower*, Europeans had been visiting New England for more than a hundred years. English, French, Italian, Spanish, and Portuguese mariners regularly plied the coastline, trading what they could, occasionally kidnapping the inhabitants for slaves. New England, the Europeans saw, was thickly settled and well defended. In 1605 and 1606 Samuel de Champlain visited Cape Cod, hoping to establish a French base. He abandoned the idea. Too many people already lived there. A year later Sir Ferdinando Gorges—British despite his name—tried to establish an English community in southern Maine. It had more founders than Plymouth and seems to have been better organized. Confronted by numerous well-armed local Indians, the settlers abandoned the project within months. The Indians at Plymouth would surely have been an equal obstacle to my ancestor and his ramshackle expedition had disease not intervened.

FACED WITH such stories, historians have long wondered how many people lived in the Americas at the time of contact. "Debated since Columbus attempted a partial census on Hispaniola in 1496," William Denevan has written, this "remains one of the great inquiries of history." (In 1976 Denevan assembled and edited an entire book on the subject, *The Native Population of the Americas in 1492*.) The first scholarly estimate of the indigenous population was made in 1910 by James Mooney, a distinguished ethnographer at the Smithsonian Institution. Combing through old documents, he concluded that in 1491 North America had 1.15 million inhabitants. Mooney's glittering reputation ensured that most subsequent researchers accepted his figure uncritically.

That changed in 1966, when Henry F. Dobyns published "Estimating Aboriginal American Population: An Appraisal of Techniques with a New Hemispheric Estimate," in the journal *Current Anthropology*. Despite the carefully neutral title, his argument was thunderous, its impact long-lasting. In the view of James Wilson, the author of *The Earth Shall Weep* (1998), a history of indigenous Americans, Dobyns's colleagues "are still struggling to get out of the crater that paper left in anthropology." Not only anthropologists were affected. Dobyns's estimate proved to be one of the opening rounds in today's culture wars.

Dobyns began his exploration of pre-Columbian Indian demography in the early 1950s, when he was a graduate student. At the invitation of a friend, he spent a few months in northern Mexico, which is full of Spanish-era missions. There he poked through the crumbling leather-bound ledgers in which Jesuits recorded local births and deaths. Right away he noticed how many more deaths there were. The Spaniards arrived, and then Indians died—in huge numbers, at incredible rates. It hit him, Dobyns told me recently, "like a club right between the eyes."

It took Dobyns eleven years to obtain his Ph.D. Along the way he joined a rural-development project in Peru, which until colonial times was the seat of the Incan empire. Remembering what he had seen at the northern fringe of the Spanish conquest, Dobyns decided to compare it with figures for the south. He burrowed into the papers of the Lima cathedral and read apologetic Spanish histories. The Indians in Peru, Dobyns concluded, had faced plagues from the day the conquistadors showed up—in fact, before then: smallpox arrived around 1525, seven years ahead of the Spanish. Brought to Mexico apparently by a single sick Spaniard, it swept south and eliminated more than half the population of the Incan empire. Smallpox claimed the Incan dictator Huayna Capac and much of his family, setting off a calamitous war of succession. So complete was the chaos that Francisco Pizarro was able to seize an empire the size of Spain and Italy combined with a force of 168 men.

Smallpox was only the first epidemic. Typhus (probably) in 1546, influenza and smallpox together in 1558, smallpox again in 1589, diphtheria in 1614, measles in 1618—all ravaged the remains of Incan culture. Dobyns was the first social scientist to piece together this awful picture, and he naturally rushed his findings into print. Hardly anyone paid attention. But Dobyns was already working on a second, related question: If all those people died, how many had been living there to begin with? Before Columbus, Dobyns calculated, the Western Hemisphere held 90 to 112 million people. Another way of saying this is that in 1491 more people lived in the Americas than in Europe.

His argument was simple but horrific. It is well known that Native Americans had no experience with many European diseases and were therefore immunologically unprepared—"virgin soil," in the metaphor of epidemiologists. What Dobyns realized was that such diseases could have swept from the coastlines initially visited by Europeans to inland areas controlled by Indians who had never seen a white person. The first whites to explore many parts of the Americas may therefore have encountered places that were already depopulated. Indeed, Dobyns argued, they must have done so.

Peru was one example, the Pacific Northwest another. In 1792 the British navigator George Vancouver led the first European expedition to survey Puget Sound. He found a vast charnel house: human remains "promiscuously scattered about the beach, in great numbers." Smallpox, Vancouver's crew discovered, had preceded them. Its few survivors, second lieutenant Peter Puget noted, were "most terribly pitted . . . indeed many have lost their Eyes." In *Pox Americana* (2001), Elizabeth Fenn, a historian at George Washington University, contends that the disaster on the northwest coast was but a small part of a continental pandemic that erupted near Boston in 1774 and cut down Indians from Mexico to Alaska.

Because smallpox was not endemic in the Americas, colonials, too, had not acquired any immunity. The virus, an equal-opportunity killer, swept through the Continental Army and stopped the drive into Quebec. The American Revolution would be lost, Washington and other rebel leaders feared, if the contagion did to the colonists what it had done to the Indians. "The small Pox! The small Pox!" John Adams wrote to his wife, Abigail. "What shall We do with it?" In retrospect, Fenn says, "One of George Washington's most brilliant moves was to inoculate the army against smallpox during the Valley Forge winter of '78." Without inoculation smallpox could easily have given the United States back to the British.

So many epidemics occurred in the Americas, Dobyns argued, that the old data used by Mooney and his successors represented population nadirs. From the few cases in which before-and-after totals are known with relative certainty, Dobyns estimated that in the first 130 years of contact about 95 percent of the people in the Americas died—the worst demographic calamity in recorded history.

Dobyns's ideas were quickly attacked as politically motivated, a push from the hate-America crowd to inflate the toll of imperialism. The attacks continue to this day. "No question about it, some people want those higher numbers," says Shepard Krech III, a Brown University anthropologist who is the author of The Ecological Indian (1999). These people, he says, were thrilled when Dobyns revisited the subject in a book, Their Numbers Become Thinned (1983)—and revised his own estimates upward. Perhaps Dobyns's most vehement critic is David Henige, a bibliographer of Africana at the University of Wisconsin, whose Numbers from Nowhere (1998) is a landmark in the literature of demographic fulmination. "Suspect in 1966, it is no less suspect nowadays," Henige wrote of Dobyns's work. "If anything, it is worse."

When Henige wrote Numbers from Nowhere, the fight about pre-Columbian populations had already consumed forests' worth of trees; his bibliography is ninety pages long. And the dispute shows no sign of abating. More and more people have jumped in. This is partly because the subject is inherently fascinating. But more likely the increased interest in the debate is due to the growing realization of the high political and ecological stakes.

Inventing by the Millions

ON MAY 30, 1539, Hernando de Soto landed his private army near Tampa Bay, in Florida. Soto, as he was called, was a novel figure: half warrior, half venture capitalist. He had grown very rich very young by becoming a market leader in the nascent trade for Indian slaves. The profits had helped to fund Pizarro's seizure of the Incan empire, which had made Soto wealthier still. Looking quite literally for new worlds to conquer, he persuaded the Spanish Crown to let him loose in North America. He spent one fortune to make another. He came to Florida with 200 horses, 600 soldiers, and 300 pigs.

From today's perspective, it is difficult to imagine the ethical system that would justify Soto's actions. For four years his force, looking for gold, wandered through what is now Florida, Georgia, North and South Carolina, Tennessee, Alabama, Mississippi, Arkansas, and Texas, wrecking almost everything it touched. The inhabitants often fought back vigorously, but they had never before encountered an army with horses and guns. Soto died of fever with his expedition in ruins; along the way his men had managed to rape, torture, enslave, and kill countless Indians. But the worst thing the Spaniards did, some researchers say, was entirely without malice—bring the pigs.

According to Charles Hudson, an anthropologist at the University of Georgia who spent fifteen years reconstructing the path of the expedition, Soto crossed the Mississippi a few miles downstream from the present site of Memphis. It was a nervous passage: the Spaniards were watched by several thousand Indian warriors. Utterly without fear, Soto brushed past the Indian force into what is now eastern Arkansas, through thickly settled land—"very well peopled with large towns," one of his men later recalled, "two or three of which were to be seen from one town." Eventually the Spaniards approached a cluster of small cities, each protected by earthen walls, sizeable moats, and deadeye archers. In his usual fashion, Soto brazenly marched in, stole food, and marched out.

After Soto left, no Europeans visited this part of the Mississippi Valley for more than a century. Early in 1682 whites appeared again, this time Frenchmen in canoes. One of them was Réné-Robert Cavelier, Sieur de La Salle. The French passed through the area where Soto had found cities cheek by jowl. It was deserted—La Salle didn't see an Indian village for 200 miles. About fifty settlements existed in this strip of the Mississippi when Soto showed up, according to Anne Ramenofsky, an anthropologist at the University of New Mexico. By La Salle's time the number had shrunk to perhaps ten, some probably inhabited by recent immigrants. Soto "had a privileged glimpse" of an Indian world, Hudson says. "The window opened and slammed shut. When the French came in and the record opened up again, it was a transformed reality. A civilization crumbled. The question is, how did this happen?"

The question is even more complex than it may seem. Disaster of this magnitude suggests epidemic disease. In the view of Ramenofsky and Patricia Galloway, an anthropologist at the University of Texas, the source of the contagion was very likely not Soto's army but its ambulatory meat locker: his 300 pigs. Soto's force itself was too small to be an effective biological weapon. Sicknesses like measles and smallpox would have burned through his 600 soldiers long before they reached the Mississippi. But the same would not have held true for the pigs, which multiplied rapidly and were able to transmit their diseases to wildlife in the surrounding forest. When human beings and domesticated animals live close together, they trade microbes with abandon. Over time mutation spawns new diseases: avian influenza becomes human influenza, bovine rinderpest becomes measles. Unlike Europeans, Indians did not live in close quarters with animals—they domesticated only the dog, the llama, the alpaca, the guinea pig, and, here and there, the turkey and the Muscovy duck. In some ways this is not surprising: the New World had fewer animal candidates for taming than the Old. Moreover, few Indians carry the gene that permits adults to digest lactose, a form of sugar abundant in milk. Non-milk-drinkers, one imagines, would be less likely to work at domesticating milk-giving animals. But this is guesswork. The fact is that what scientists call zoonotic disease was little known in the Americas. Swine alone can disseminate anthrax, brucellosis, leptospirosis, taeniasis, trichinosis, and tuberculosis. Pigs breed exuberantly and can transmit diseases to deer and turkeys. Only a few of Soto's pigs would have had to wander off to infect the forest.

Indeed, the calamity wrought by Soto apparently extended across the whole Southeast. The Coosa city-states, in western Georgia, and the Caddoan-speaking civilization, centered on the Texas-Arkansas border, disintegrated soon after Soto appeared. The Caddo had had a taste for monumental architecture: public plazas, ceremonial platforms, mausoleums. After Soto's army left, notes Timothy K. Perttula, an archaeological consultant in Austin, Texas, the Caddo stopped building community centers and began digging community cemeteries. Between Soto's and La Salle's visits, Perttula believes, the Caddoan population fell from about 200,000 to about 8,500—a drop of nearly 96 percent. In the eighteenth century the tally shrank further, to 1,400. An equivalent loss today in the population of New York City would reduce it to 56,000—not enough to fill Yankee Stadium. "That's one reason whites think of Indians as nomadic hunters," says Russell Thornton, an anthropologist at the University of California at Los Angeles. "Everything else—all the heavily populated urbanized societies—was wiped out."

Could a few pigs truly wreak this much destruction? Such apocalyptic scenarios invite skepticism. As a rule, viruses, microbes, and parasites are rarely lethal on so wide a scale—a pest that wipes out its host species does not have a bright evolutionary future. In its worst outbreak, from 1347 to 1351, the European Black Death claimed only a third of its victims. (The rest survived, though they were often disfigured or crippled by its effects.) The Indians in Soto's path, if Dobyns, Ramenofsky, and Perttula are correct, endured losses that were incomprehensibly greater.

One reason is that Indians were fresh territory for many plagues, not just one. Smallpox, typhoid, bubonic plague, influenza, mumps, measles, whooping cough—all rained down on the Americas in the century after Columbus. (Cholera, malaria, and scarlet fever came later.) Having little experience with epidemic diseases, Indians had no knowledge of how to combat them. In contrast, Europeans were well versed in the brutal logic of quarantine. They boarded up houses in which plague appeared and fled to the countryside. In Indian New England, Neal Salisbury, a historian at Smith College, wrote in *Manitou and Providence* (1982), family and friends gathered with the shaman at the sufferer's bedside to wait out the illness—a practice that "could only have served to spread the disease more rapidly."

Indigenous biochemistry may also have played a role. The immune system constantly scans the body for molecules that it can recognize as foreign—molecules belonging to an invading virus, for instance. No one's immune system can identify all foreign presences. Roughly speaking, an individual's set of defensive tools is known as his MHC type. Because many bacteria and viruses mutate easily, they usually attack in the form of several slightly different strains. Pathogens win when MHC types miss some of the strains and the immune system is not stimulated to act. Most human groups contain many MHC types; a strain that slips by one person's defenses will be nailed by the defenses of the next. But, according to Francis L. Black, an epidemiologist at Yale University, Indians are characterized by unusually homogenous MHC types. One out of three South American Indians have similar MHC types; among Africans the corresponding figure is one in 200. The cause is a matter for Darwinian speculation, the effects less so.

In 1966 Dobyns's insistence on the role of disease was a shock to his colleagues. Today the impact of European pathogens on the New World is almost undisputed. Nonetheless, the fight over Indian numbers continues with undiminished fervor. Estimates of the population of North America in 1491 disagree by an order of magnitude—from 18 million, Dobyns's revised figure, to 1.8 million, calculated by Douglas H. Ubelaker, an anthropologist at the Smithsonian. To some "high counters," as David Henige calls them, the low counters' refusal to relinquish the vision of an empty continent is irrational or worse. "Non-Indian 'experts' always want to minimize the size of aboriginal populations," says Lenore Stiffarm, a Native American–education specialist at the University of Saskatchewan. The smaller the numbers of Indians, she believes, the easier it is to regard the continent as having been up for grabs. "It's perfectly acceptable to move into unoccupied land," Stiffarm says. "And land with only a few 'savages' is the next best thing."

"Most of the arguments for the very large numbers have been theoretical," Ubelaker says in defense of low counters. "When you try to marry the theoretical arguments to the data that are available on individual groups in different regions, it's hard to find support for those numbers." Archaeologists, he says, keep searching for the settlements in which those millions of people supposedly lived, with little success. "As more and more excavation is done, one would expect to see more evidence for dense populations than has thus far emerged." Dean Snow, the Pennsylvania State anthropologist, examined Colonial-era Mohawk Iroquois sites and found "no support for the notion that ubiquitous pandemics swept the region." In his view, asserting that the continent was filled with people who left no trace is like looking at an empty bank account and claiming that it must once have held millions of dollars.

The low counters are also troubled by the Dobynsian procedure for recovering original population numbers: applying an assumed death rate, usually 95 percent, to the observed population nadir. Ubelaker believes that the lowest point for Indians in North America was around 1900, when their numbers fell to about half a million. Assuming a 95 percent death rate, the pre-contact population would have been 10 million. Go up one percent, to a 96 percent death rate, and the figure jumps to 12.5 million—arithmetically creating more than two million people from a tiny increase in mortality rates. At 98 percent the number bounds to 25 million. Minute changes in baseline assumptions produce wildly different results.

"It's an absolutely unanswerable question on which tens of thousands of words have been spent to no purpose," Henige says. In 1976 he sat in on a seminar by William Denevan, the Wisconsin geographer. An "epiphanic moment" occurred when he read shortly afterward that scholars had "uncovered" the existence of eight million people in Hispaniola. *Can you just invent millions of people?* he wondered. "We can make of the historical record that there was depopulation and movement of people from internecine warfare and diseases," he says. "But as for how much, who knows? When we start putting numbers to something like that—applying large figures like ninety-five percent—we're saying things we shouldn't say. The number implies a level of knowledge that's impossible."

Nonetheless, one must try—or so Denevan believes. In his estimation the high counters (though not the highest counters) seem to be winning the argument, at least for now. No definitive data exist, he says, but the majority of the extant evidentiary scraps support their side. Even Henige is no low counter. When I asked him what he thought the population of the Americas was before Columbus, he insisted that any answer would be speculation and made me promise not to print what he was going to say next. Then he named a figure that forty years ago would have caused a commotion.

To Elizabeth Fenn, the smallpox historian, the squabble over numbers obscures a central fact. Whether one million or 10 million or 100 million died, she believes, the pall of sorrow that engulfed the hemisphere was immeasurable. Languages, prayers, hopes, habits, and dreams—entire ways of life hissed away like steam. The Spanish and the Portuguese lacked the germ theory of disease and could not explain what was happening (let alone stop it). Nor can we explain it; the ruin was too long ago and too all-encompassing. In the long run, Fenn says, the consequential finding is not that many people died but that many people once lived. The Americas were filled with a stunningly diverse assortment of peoples who had knocked about the continents for millennia. "You have to wonder," Fenn says. "What were all those people *up* to in all that time?"

Buffalo Farm

IN 1810 HENRY BRACKENRIDGE came to Cahokia, in what is now southwest Illinois, just across the Mississippi from St. Louis. Born close to the frontier, Brackenridge was a budding adventure writer; his *Views of Louisiana,* published three years later, was a kind of nineteenth-century *Into Thin Air,* with terrific adventure but without tragedy. Brackenridge had an eye for archaeology, and he had heard that Cahokia was worth a visit. When he got there, trudging along the desolate Cahokia River, he was "struck with a degree of astonishment." Rising from the muddy bottomland was a "stupendous pile of earth," vaster than the Great Pyramid at Giza. Around it were more than a hundred smaller mounds, covering an area of five square miles. At the time, the area was almost uninhabited. One can only imagine what passed through Brackenridge's mind as he walked alone to the ruins of the biggest Indian city north of the Rio Grande.

To Brackenridge, it seemed clear that Cahokia and the many other ruins in the Midwest had been constructed by Indians. It was not so clear to everyone else. Nineteenth-century writers attributed them to, among others, the Vikings, the Chinese, the "Hindoos," the ancient Greeks, the ancient Egyptians, lost tribes of Israelites, and even straying bands of Welsh. (This last claim was surprisingly widespread; when Lewis and Clark surveyed the Missouri, Jefferson told them to keep an eye out for errant bands of Welsh-speaking white Indians.) The historian George Bancroft, dean of his profession, was a dissenter: the earthworks, he wrote in 1840, were purely natural formations.

Bancroft changed his mind about Cahokia, but not about Indians. To the end of his days he regarded them as "feeble barbarians, destitute of commerce and of political connection." His characterization lasted, largely unchanged, for more than a century. Samuel Eliot Morison, the winner of two Pulitzer Prizes, closed his monumental *European Discovery of America* (1974) with the observation that Native Americans expected only "short and brutish lives, void of hope for any future." As late as 1987 *American History: A Survey*, a standard high school textbook by three well-known historians, described the Americas before Columbus as "empty of mankind and its works." The story of Europeans in the New World, the book explained, "is the story of the creation of a civilization where none existed."

Alfred Crosby, a historian at the University of Texas, came to other conclusions. Crosby's *The Columbian Exchange: Biological Consequences of 1492* caused almost as much of a stir when it was published, in 1972, as Henry Dobyns's calculation of Indian numbers six years earlier, though in different circles. Crosby was a standard names-and-battles historian who became frustrated by the random contingency of political events. "Some trivial thing happens and you have this guy winning the presidency instead of that guy," he says. He decided to go deeper. After he finished his manuscript, it sat on his shelf—he couldn't find a publisher willing to be associated with his new ideas. It took him three years to persuade a small editorial house to put it out. *The Columbian Exchange* has been in print ever since; a companion, *Ecological Imperialism: The Biological Expansion of Europe, 900–1900*, appeared in 1986.

Human history, in Crosby's interpretation, is marked by two world-altering centers of invention: the Middle East and central Mexico, where Indian groups independently created nearly all of the Neolithic innovations, writing included. The Neolithic Revolution began in the Middle East about 10,000 years ago. In the next few millennia humankind invented the wheel, the metal tool, and agriculture. The Sumerians eventually put these inventions together, added writing, and became the world's first civilization. Afterward Sumeria's heirs in Europe and Asia frantically copied one another's happiest discoveries; innovations ricocheted from one corner of Eurasia to another, stimulating technological progress. Native Americans, who had crossed to Alaska before Sumeria, missed out on the bounty. "They had to do everything on their own," Crosby says. Remarkably, they succeeded.

When Columbus appeared in the Caribbean, the descendants of the world's two Neolithic civilizations collided, with overwhelming consequences for both. American Neolithic development occurred later than that of the Middle East, possibly because the Indians needed more time to build up the requisite population density. Without beasts of burden they could not capitalize on the wheel (for individual workers on uneven terrain skids are nearly as effective as carts for hauling), and they never developed steel. But in agriculture they handily outstripped the children of Sumeria. Every tomato in Italy, every potato in Ireland, and every hot pepper in Thailand came from this hemisphere. Worldwide, more than half the crops grown today were initially developed in the Americas.

Maize, as corn is called in the rest of the world, was a triumph with global implications. Indians developed an extraordinary number of maize varieties for different growing conditions, which meant that the crop could and did spread throughout the planet. Central and Southern Europeans became particularly dependent on it; maize was the staple of Serbia, Romania, and Moldavia by the nineteenth century. Indian crops dramatically reduced hunger, Crosby says, which led to an Old World population boom.

Along with peanuts and manioc, maize came to Africa and transformed agriculture there, too. "The probability is that the population of Africa was greatly increased because of maize and other American Indian crops," Crosby says. "Those extra people helped make the slave trade possible." Maize conquered Africa at the time when introduced diseases were leveling Indian societies. The Spanish, the Portuguese, and the British were alarmed by the death rate among Indians, because they wanted to exploit them as workers. Faced with a labor shortage, the Europeans turned their eyes to Africa. The continent's quarrelsome societies helped slave traders to siphon off millions of people. The maize-fed population boom, Crosby believes, let the awful trade continue without pumping the well dry.

Back home in the Americas, Indian agriculture long sustained some of the world's largest cities. The Aztec capital of Tenochtitlán dazzled Hernán Cortés in 1519; it was bigger than Paris, Europe's greatest metropolis. The Spaniards gawped like hayseeds at the wide streets, ornately carved buildings, and markets bright with goods from hundreds of miles away. They had never before seen a city with botanical gardens, for the excellent reason that none existed in Europe. The same novelty attended the force of a thousand men that kept the crowded streets immaculate. (Streets that weren't ankle-deep in sewage! The conquistadors had never heard of such a thing.) Central America was not the only locus of prosperity. Thousands of miles north, John Smith, of Pocahontas fame, visited Massachusetts in 1614, before it was emptied by disease, and declared that the land was "so planted with Gardens and Corne fields, and so well inhabited with a goodly, strong and well proportioned people . . . [that] I would rather live here than any where."

Smith was promoting colonization, and so had reason to exaggerate. But he also knew the hunger, sickness, and oppression of European life. France—"by any standards a privileged country," according to its great historian, Fernand Braudel—experienced seven nationwide famines in the fifteenth century and thirteen in the sixteenth. Disease was hunger's constant companion. During epidemics in London the dead were heaped onto carts "like common dung" (the simile is Daniel Defoe's) and trundled through the streets. The infant death rate in London orphanages, according to one contemporary source, was 88 percent. Governments were harsh, the rule of law arbitrary. The gibbets poking up in the background of so many old paintings were, Braudel observed, "merely a realistic detail."

The Earth Shall Weep, James Wilson's history of Indian America, puts the comparison bluntly: "the western hemisphere was larger, richer, and more populous than Europe." Much of it was freer, too. Europeans, accustomed to the serfdom that thrived from Naples to the Baltic Sea, were puzzled and alarmed by the democratic spirit and respect for human rights in many Indian societies, especially those in North America. In theory, the sachems of New England Indian groups were absolute monarchs. In practice, the colonial leader Roger Williams wrote, "they will not conclude of aught . . . unto which the people are averse."

Pre-1492 America wasn't a disease-free paradise, Dobyns says, although in his "exuberance as a writer," he told me recently, he once made that claim. Indians had ailments of their own, notably parasites, tuberculosis, and anemia. The daily grind was wearing; life-spans in America were only as long as or a little longer than those in Europe, if the evidence of indigenous graveyards is to be believed. Nor was it a political utopia—the Inca, for instance, invented refinements to totalitarian rule that would have intrigued Stalin. Inveterate practitioners of what the historian Francis Jennings described as "state terrorism practiced horrifically on a huge scale," the Inca ruled so cruelly that one can speculate that their surviving subjects might actually have been better off under Spanish rule.

I asked seven anthropologists, archaeologists, and historians if they would rather have been a typical Indian or a typical European in 1491. None was delighted by the question, because it required judging the past by the standards of today—a fallacy disparaged as "presentism" by social scientists. But every one chose to be an Indian. Some early colonists gave the same answer. Horrifying the leaders of Jamestown and Plymouth, scores of English ran off to live with the Indians. My ancestor shared their desire, which is what led to the trumped-up murder charges against him—or that's what my grandfather told me, anyway.

As for the Indians, evidence suggests that they often viewed Europeans with disdain. The Hurons, a chagrined missionary reported, thought the French possessed "little intelligence in comparison to themselves." Europeans, Indians said, were physically weak, sexually untrustworthy, atrociously ugly, and just plain dirty. (Spaniards, who seldom if ever bathed, were amazed by the Aztec desire for personal cleanliness.) A Jesuit reported that the "Savages" were disgusted by handkerchiefs: "They say, we place what is unclean in a fine white piece of linen, and put it away in our pockets as something very precious, while they throw it upon the ground." The Micmac scoffed at the notion of French superiority. If Christian civilization was so wonderful, why were its inhabitants leaving?

Like people everywhere, Indians survived by cleverly exploiting their environment. Europeans tended to manage land by breaking it into fragments for farmers and herders. Indians often worked on such a grand scale that the scope of their ambition can be hard to grasp. They created small plots, as Europeans did (about 1.5 million acres of terraces still exist in the Peruvian Andes), but they also reshaped entire landscapes to suit their purposes. A principal tool was fire, used to keep down underbrush and create the open, grassy conditions favorable for game. Rather than domesticating animals for meat, Indians retooled whole ecosystems to grow bumper crops of elk, deer, and bison. The first white settlers in Ohio found forests as open as English parks—they could drive carriages through the woods. Along the Hudson River the annual fall burning lit up the banks for miles on end; so flashy was the show that the Dutch in New Amsterdam boated upriver to goggle at the blaze like children at fireworks. In North America, Indian torches had their biggest impact on the Midwestern prairie, much or most of which was created and maintained by fire. Millennia of exuberant burning shaped the plains into vast buffalo farms. When Indian societies disintegrated, forest invaded savannah in Wisconsin, Illinois, Kansas, Nebraska, and the Texas Hill Country. Is it possible that the Indians changed the Americas more than the invading Europeans did? "The answer is probably yes for most regions for the next 250 years or so" after Columbus, William Denevan wrote, "and for some regions right up to the present time."

When scholars first began increasing their estimates of the ecological impact of Indian civilization, they met with considerable resistance from anthropologists and archaeologists. Over time the consensus in the human sciences changed. Under Denevan's direction, Oxford University Press has just issued the third volume of a huge catalogue of the "cultivated landscapes" of the Americas. This sort of phrase still provokes vehement objection—but the main dissenters are now ecologists and environmentalists. The disagreement is encapsulated by Amazonia, which has become *the* emblem of vanishing wilderness—an admonitory image of untouched Nature. Yet recently a growing number of researchers have come to believe that Indian societies had an enormous environmental impact on the jungle. Indeed, some anthropologists have called the Amazon forest itself a cultural artifact—that is, an artificial object.

Green Prisons

NORTHERN VISITORS' first reaction to the storied Amazon rain forest is often disappointment. Ecotourist brochures evoke the immensity of Amazonia but rarely dwell on its extreme flatness. In the river's first 2,900 miles the vertical drop is only 500 feet. The river oozes like a huge runnel of dirty metal through a landscape utterly devoid of the romantic crags, arroyos, and heights that signify wildness and natural spectacle to most North Americans. Even the animals are invisible, although sometimes one can hear the bellow of monkey choruses. To the untutored eye—mine, for instance—the forest seems to stretch out in a monstrous green tangle as flat and incomprehensible as a printed circuit board.

The area east of the lower-Amazon town of Santarém is an exception. A series of sandstone ridges several hundred feet high reach down from the north, halting almost at the water's edge. Their tops stand drunkenly above the jungle like old tombstones. Many of the caves in the buttes are splattered with ancient petroglyphs—renditions of hands, stars, frogs, and human figures, all reminiscent of Miró, in overlapping red and yellow and brown. In recent years one of these caves, La Caverna da Pedra Pintada (Painted Rock Cave), has drawn attention in archaeological circles.

Wide and shallow and well lit, Painted Rock Cave is less thronged with bats than some of the other caves. The arched entrance is twenty feet high and lined with rock paintings. Out front is a sunny natural patio suitable for picnicking, edged by a few big rocks. People lived in this cave more than 11,000 years ago. They had no agriculture yet, and instead ate fish and fruit and built fires. During a recent visit I ate a sandwich atop a particularly inviting rock and looked over the forest below. The first Amazonians, I thought, must have done more or less the same thing.

In college I took an introductory anthropology class in which I read *Amazonia: Man and Culture in a Counterfeit Paradise* (1971), perhaps the most influential book ever written about the Amazon, and one that deeply impressed me at the time. Written by Betty J. Meggers, the Smithsonian archaeologist, *Amazonia* says that the apparent lushness of the rain forest is a sham. The soils are poor and can't hold nutrients—the jungle flora exists only because it snatches up everything worthwhile before it leaches away in the rain. Agriculture, which depends on extracting the wealth of the soil, therefore faces inherent ecological limitations in the wet desert of Amazonia.

As a result, Meggers argued, Indian villages were forced to remain small— any report of "more than a few hundred" people in permanent settlements, she told me recently, "makes my alarm bells go off." Bigger, more complex societies would inevitably overtax the forest soils, laying waste to their own foundations. Beginning in 1948 Meggers and her late husband, Clifford Evans, excavated a chiefdom on Marajó, an island twice the size of New Jersey that sits like a gigantic stopper in the mouth of the Amazon. The Marajóara, they concluded, were failed offshoots of a sophisticated culture in the Andes. Transplanted to the lush trap of the Amazon, the culture choked and died.

Green activists saw the implication: development in tropical forests destroys both the forests and their developers. Meggers's account had enormous public impact—*Amazonia* is one of the wellsprings of the campaign to save rain forests.

Then Anna C. Roosevelt, the curator of archaeology at Chicago's Field Museum of Natural History, re-excavated Marajó. Her complete report, *Moundbuilders of the Amazon* (1991), was like the anti-matter version of *Amazonia*. Marajó, she argued, was "one of the outstanding indigenous cultural achievements of the New World," a powerhouse that lasted for more than a thousand years, had "possibly well over 100,000" inhabitants, and covered thousands of square miles. Rather than damaging the forest, Marajó's "earth construction" and "large, dense populations" had *improved* it: the most luxuriant and diverse growth was on the mounds formerly occupied by the Marajóara. "If you listened to Meggers's theory, these places should have been ruined," Roosevelt says.

Meggers scoffed at Roosevelt's "extravagant claims," "polemical tone," and "defamatory remarks." Roosevelt, Meggers argued, had committed the beginner's error of mistaking a site that had been occupied many times by small, unstable groups for a single, long-lasting society. "[Archaeological remains] build up on areas of half a kilometer or so," she told me, "because [shifting Indian groups] don't land exactly on the same spot. The decorated types of pottery don't change much over time, so you can pick up a bunch of chips and say, 'Oh, look, it was all one big site!' Unless you know what you're doing, of course." Centuries after the conquistadors, "the myth of El Dorado is being revived by archaeologists," Meggers wrote last fall in the journal *Latin American Antiquity*, referring to the persistent Spanish delusion that cities of gold existed in the jungle.

The dispute grew bitter and personal; inevitable in a contemporary academic context, it has featured vituperative references to colonialism, elitism, and employment by the CIA. Meanwhile, Roosevelt's team investigated Painted Rock Cave. On the floor of the cave what looked to me like nothing in particular turned out to be an ancient midden: a refuse heap. The archaeologists slowly scraped away sediment, traveling backward in time with every inch. When the traces of human occupation vanished, they kept digging. ("You always go a meter past sterile," Roosevelt says.) A few inches below they struck the charcoal-rich dirt that signifies human habitation—a culture, Roosevelt said later, that wasn't supposed to be there.

For many millennia the cave's inhabitants hunted and gathered for food. But by about 4,000 years ago they were growing crops—perhaps as many as 140 of them, according to Charles R. Clement, an anthropological botanist at the Brazilian National Institute for Amazonian Research. Unlike Europeans, who planted mainly annual crops, the Indians, he says, centered their agriculture on the Amazon's unbelievably diverse assortment of trees: fruits, nuts, and palms. "It's tremendously difficult to clear fields with stone tools," Clement says. "If you can plant trees, you get twenty years of productivity out of your work instead of two or three."

Planting their orchards, the first Amazonians transformed large swaths of the river basin into something more pleasing to human beings. In a widely cited article from 1989, William Balée, the Tulane anthropologist, cautiously estimated that about 12 percent of the nonflooded Amazon forest was of anthropogenic origin—directly or indirectly created by human beings. In some circles this is now seen as a conservative position. "I basically think it's all human-created," Clement told me in Brazil. He argues that Indians changed the assortment and density of species throughout the region. So does Clark Erickson, the University of Pennsylvania archaeologist, who told me in Bolivia that the lowland tropical forests of South America are among the finest works of art on the planet. "Some of my colleagues would say that's pretty radical," he said, smiling mischievously. According to Peter Stahl, an anthropologist at the State University of New York at Binghamton, "lots" of botanists believe that "what the eco-imagery would like to picture as a pristine, untouched Urwelt [primeval world] in fact has been managed by people for millennia." The phrase "built environment," Erickson says, "applies to most, if not all, Neotropical landscapes."

"Landscape" in this case is meant exactly—Amazonian Indians literally created the ground beneath their feet. According to William I. Woods, a soil geographer at Southern Illinois University, ecologists' claims about terrible Amazonian land were based on very little data. In the late 1990s Woods and others began careful measurements in the lower Amazon. They indeed found lots of inhospitable terrain. But they also discovered swaths of *terra preta*—rich, fertile "black earth" that anthropologists increasingly believe was created by human beings.

Terra preta, Woods guesses, covers at least 10 percent of Amazonia, an area the size of France. It has amazing properties, he says. Tropical rain doesn't leach nutrients from *terra preta* fields; instead the soil, so to speak, fights back. Not far from Painted Rock Cave is a 300-acre area with a two-foot layer of *terra preta* quarried by locals for potting soil. The bottom third of the layer is never removed, workers there explain, because over time it will re-create the original soil layer in its initial thickness. The reason, scientists suspect, is that *terra preta* is generated by a special suite of microorganisms that resists depletion. "Apparently," Woods and the Wisconsin geographer Joseph M. McCann argued in a presentation the summer of 2001, "at some threshold level . . . dark earth attains the capacity to perpetuate—even *regenerate* itself—thus behaving more like a living 'super'-organism than an inert material."

In as yet unpublished research the archaeologists Eduardo Neves, of the University of São Paulo; Michael Heckenberger, of the University of Florida; and their colleagues examined *terra preta* in the upper Xingu, a huge southern tributary of the Amazon. Not all Xingu cultures left behind this living earth, they discovered. But the ones that did generated it rapidly—suggesting to Woods that *terra preta* was created deliberately. In a process reminiscent of dropping microorganism-rich starter into plain dough to create sourdough bread, Amazonian peoples, he believes, inoculated bad soil with a transforming bacterial charge. Not every group of Indians there did this, but quite a few did, and over an extended period of time.

When Woods told me this, I was so amazed that I almost dropped the phone. I ceased to be articulate for a moment and said things like "wow" and "gosh." Woods chuckled at my reaction, probably because he understood what was passing through my mind. Faced with an ecological problem, I was thinking, the Indians *fixed* it. They were in the process of terraforming the Amazon when Columbus showed up and ruined everything.

Scientists should study the microorganisms in *terra preta*, Woods told me, to find out how they work. If that could be learned, maybe some version of Amazonian dark earth could be used to improve the vast expanses of bad soil that cripple agriculture in Africa—a final gift from the people who brought us tomatoes, corn, and the immense grasslands of the Great Plains.

"Betty Meggers would just die if she heard me saying this," Woods told me. "Deep down her fear is that this data will be misused." Indeed, Meggers's 2001 *Latin American Antiquity* article charged that archaeologists who say the Amazon can support agriculture are effectively telling "developers [that they] are entitled to operate without restraint." Resuscitating the myth of El Dorado, in her view, "makes us accomplices in the accelerating pace of environmental degradation." Doubtless there is something to this—although, as some of her critics responded in the same issue of the journal, it is difficult to imagine greedy plutocrats "perusing the pages of *Latin American Antiquity* before deciding to rev up the chain saws." But the new picture doesn't automatically legitimize paving the forest. Instead it suggests that for a long time big chunks of Amazonia were used nondestructively by clever people who knew tricks we have yet to learn.

I visited Painted Rock Cave during the river's annual flood, when it wells up over its banks and creeps inland for miles. Farmers in the floodplain build houses and barns on stilts and watch pink dolphins sport from their doorsteps. Ecotourists take shortcuts by driving motorboats through the drowned forest. Guys in dories chase after them, trying to sell sacks of incredibly good fruit.

All of this is described as "wilderness" in the tourist brochures. It's not, if researchers like Roosevelt are correct. Indeed, they believe that fewer people may be living there now than in 1491. Yet when my boat glided into the trees, the forest shut out the sky like the closing of an umbrella. Within a few hundred yards the human presence seemed to vanish. I felt alone and small, but in a way that was curiously like feeling exalted. If that place was not wilderness, how should I think of it? Since the fate of the forest is in our hands, what should be our goal for its future?

Novel Shores

HERNANDO DE SOTO'S EXPEDITION stomped through the Southeast for four years and apparently never saw bison. More than a century later, when French explorers came down the Mississippi, they saw "a solitude unrelieved by the faintest trace of man," the nineteenth-century historian Francis Parkman wrote. Instead the French encountered bison, "grazing in herds on the great prairies which then bordered the river."

To Charles Kay, the reason for the buffalo's sudden emergence is obvious. Kay is a wildlife ecologist in the political-science department at Utah State University. In ecological terms, he says, the Indians were the "keystone species" of American ecosystems. A keystone species, according to the Harvard biologist Edward O. Wilson, is a species "that affects the survival and abundance of many other species." Keystone species have a disproportionate impact on their ecosystems. Removing them, Wilson adds, "results in a relatively significant shift in the composition of the [ecological] community."

When disease swept Indians from the land, Kay says, what happened was exactly that. The ecological ancien régime collapsed, and strange new phenomena emerged. In a way this is unsurprising; for better or worse, humankind is a keystone species everywhere. Among these phenomena was a population explosion in the species that the Indians had kept down by hunting. After disease killed off the Indians, Kay believes, buffalo vastly extended their range. Their numbers more than sextupled. The same occurred with elk and mule deer. "If the elk were here in great numbers all this time, the archaeological sites should be chock-full of elk bones," Kay says. "But the archaeologists will tell you the elk weren't there." On the evidence of middens the number of elk jumped about 500 years ago.

Passenger pigeons may be another example. The epitome of natural American abundance, they flew in such great masses that the first colonists were stupefied by the sight. As a boy, the explorer Henry Brackenridge saw flocks "ten miles in width, by one hundred and twenty in length." For hours the birds darkened the sky from horizon to horizon. According to Thomas Neumann, a consulting archaeologist in Lilburn, Georgia, passenger pigeons "were incredibly dumb and always roosted in vast hordes, so they were very easy to harvest." Because they were readily caught and good to eat, Neumann says, archaeological digs should find many pigeon bones in the pre-Columbian strata of Indian middens. But they aren't there. The mobs of birds in the history books, he says, were "outbreak populations—always a symptom of an extraordinarily disrupted ecological system."

Throughout eastern North America the open landscape seen by the first Europeans quickly filled in with forest. According to William Cronon, of the University of Wisconsin, later colonists began complaining about how hard it was to get around. (Eventually, of course, they stripped New England almost bare of trees.) When Europeans moved west, they were preceded by two waves: one of disease, the other of ecological disturbance. The former crested with fearsome rapidity; the latter sometimes took more than a century to quiet down. Far from destroying pristine wilderness, European settlers bloodily *cre-*

ated it. By 1800 the hemisphere was chockablock with new wilderness. If "forest primeval" means a woodland unsullied by the human presence, William Denevan has written, there was much more of it in the late eighteenth century than in the early sixteenth.

Cronon's *Changes in the Land: Indians, Colonists, and the Ecology of New England* (1983) belongs on the same shelf as works by Crosby and Dobyns. But it was not until one of his articles was excerpted in *The New York Times* in 1995 that people outside the social sciences began to understand the implications of this view of Indian history. Environmentalists and ecologists vigorously attacked the anti-wilderness scenario, which they described as infected by postmodern philosophy. A small academic brouhaha ensued, complete with hundreds of footnotes. It precipitated *Reinventing Nature?* (1995), one of the few academic critiques of postmodernist philosophy written largely by biologists. *The Great New Wilderness Debate* (1998), another lengthy book on the subject, was edited by two philosophers who earnestly identified themselves as "Euro-American men [whose] cultural legacy is patriarchal Western civilization in its current postcolonial, globally hegemonic form."

It is easy to tweak academics for opaque, self-protective language like this. Nonetheless, their concerns were quite justified. Crediting Indians with the role of keystone species has implications for the way the current Euro-American members of that keystone species manage the forests, watersheds, and endangered species of America. Because a third of the United States is owned by the federal government, the issue inevitably has political ramifications. In Amazonia, fabled storehouse of biodiversity, the stakes are global.

Guided by the pristine myth, mainstream environmentalists want to preserve as much of the world's land as possible in a putatively intact state. But "intact," if the new research is correct, means "run by human beings for human purposes." Environmentalists dislike this, because it seems to mean that anything goes. In a sense they are correct. Native Americans managed the continent as they saw fit. Modern nations must do the same. If they want to return as much of the landscape as possible to its 1491 state, they will have to find it within themselves to create the world's largest garden.

ATUL GAWANDE

The Learning Curve

FROM *THE NEW YORKER*

"Practice, practice, practice" goes the punch line to the old joke, and it's as true for surgeons as it is for musicians. Atul Gawande, who has chronicled his own surgical training with honesty and humor, shares his experiences—and anxieties—about the way doctors learn their skills: performing supposedly routine procedures on unsuspecting patients.

The patient needed a central line. "Here's your chance," S., the chief resident, said. I had never done one before. "Get set up and then page me when you're ready to start."

It was my fourth week in surgical training. The pockets of my short white coat bulged with patient printouts, laminated cards with instructions for doing CPR and reading EKGs and using the dictation system, two surgical handbooks, a stethoscope, wound-dressing supplies, meal tickets, a penlight, scissors, and about a dollar in loose change. As I headed up the stairs to the patient's floor, I rattled.

This will be good, I tried to tell myself: my first real procedure. The patient—fiftyish, stout, taciturn—was recovering from abdominal surgery he'd had about a week earlier. His bowel function hadn't yet returned, and he was unable to eat. I explained to him that he needed intravenous nutrition and that this required a "special line" that would go into his chest. I said that I would put the line in him while he was in his bed, and that it would involve my numbing a spot on his chest with a local anesthetic, and then threading the line in. I did not say that the line was eight inches long and would go into his vena cava, the main blood vessel to his heart. Nor did I say how tricky the procedure could be. There were "slight risks" involved, I said, such as bleeding and lung collapse; in experienced hands, complications of this sort occur in fewer than one case in a hundred.

But, of course, mine were not experienced hands. And the disasters I knew about weighed on my mind: the woman who had died within minutes from massive bleeding when a resident lacerated her vena cava; the man whose chest had to be opened because a resident lost hold of a wire inside the line, which then floated down to the patient's heart; the man who had a cardiac arrest when the procedure put him into ventricular fibrillation. I said nothing of such things, naturally, when I asked the patient's permission to do his line. He said, "OK."

I had seen S. do two central lines; one was the day before, and I'd attended to every step. I watched how she set out her instruments and laid her patient down and put a rolled towel between his shoulder blades to make his chest arch out. I watched how she swabbed his chest with antiseptic, injected lidocaine, which is a local anesthetic, and then, in full sterile garb, punctured his chest near his clavicle with a fat three-inch needle on a syringe. The patient hadn't even flinched. She told me how to avoid hitting the lung ("Go in at a steep angle," she'd said. "Stay *right* under the clavicle"), and how to find the subclavian vein, a branch to the vena cava lying atop the lung near its apex ("Go in at a steep angle. Stay *right* under the clavicle"). She pushed the needle in almost all the way. She drew back on the syringe. And she was in. You knew because the syringe filled with maroon blood. ("If it's bright red, you've hit an artery," she said. "That's not good.") Once you have the tip of this needle poking in the vein, you somehow have to widen the hole in the vein wall, fit the catheter in, and snake it in the right direction—down to the heart, rather than up to the brain—all without tearing through vessels, lung, or anything else.

To do this, S. explained, you start by getting a guide wire in place. She pulled the syringe off, leaving the needle in. Blood flowed out. She picked up a two-foot-long twenty-gauge wire that looked like the steel D string of an electric guitar, and passed nearly its full length through the needle's bore, into the vein, and onward toward the vena cava. "Never force it in," she warned, "and never, ever let go of it." A string of rapid heartbeats fired off on the cardiac monitor, and she quickly pulled the wire back an inch. It had poked into the heart, causing momentary fibrillation. "Guess we're in the right place," she said to me quietly. Then to the patient: "You're doing great. Only a few minutes now." She pulled the needle out over the wire and replaced it with a bullet of thick, stiff plastic, which she pushed in tight to widen the vein opening. She then removed this dilator and threaded the central line—a spaghetti-thick, flexible yellow plastic tube—over the wire until it was all the way in. Now she could remove the wire. She flushed the line with a heparin solution and sutured it to the patient's chest. And that was it.

Today, it was my turn to try. First, I had to gather supplies—a central-line kit, gloves, gown, cap, mask, lidocaine—which took me forever. When I finally had the stuff together, I stopped for a minute outside the patient's door, trying to recall the steps. They remained frustratingly hazy. But I couldn't put it off any longer. I had a page-long list of other things to get done: Mrs. A needed to be discharged; Mr. B needed an abdominal ultrasound arranged; Mrs. C needed her skin staples removed. And every fifteen minutes or so I was getting paged with more tasks: Mr. X was nauseated and needed to be seen; Miss Y's family was here and needed "someone" to talk to them; Mr. Z needed a laxative. I took a deep breath, put on my best don't-worry-I-know-what-I'm-doing look, and went in.

I placed the supplies on a bedside table, untied the patient's gown, and laid him down flat on the mattress, with his chest bare and his arms at his sides. I flipped on a fluorescent overhead light and raised his bed to my height. I paged S. I put on my gown and gloves and, on a sterile tray, laid out the central line, the guide wire, and other materials from the kit. I drew up five cc's of lidocaine in a syringe, soaked two sponge sticks in the yellow-brown Betadine, and opened up the suture packaging.

S. arrived. "What's his platelet count?"

My stomach knotted. I hadn't checked. That was bad: too low and he could have a serious bleed from the procedure. She went to check a computer. The count was acceptable.

Chastened, I started swabbing his chest with the sponge sticks. "Got the shoulder roll underneath him?" S. asked. Well, no, I had forgotten that, too. The patient gave me a look. S., saying nothing, got a towel, rolled it up, and slipped it under his back for me. I finished applying the antiseptic and then draped him so that only his right upper chest was exposed. He squirmed a bit beneath the drapes. S. now inspected my tray. I girded myself.

"Where's the extra syringe for flushing the line when it's in?" Damn. She went out and got it.

I felt for my landmarks. *Here?* I asked with my eyes, not wanting to undermine the patient's confidence any further. She nodded. I numbed the spot with lidocaine. ("You'll feel a stick and a burn now, sir.") Next, I took the three-inch needle in hand and poked it through the skin. I advanced it slowly and uncertainly, a few millimeters at a time. This is a big goddam needle, I kept thinking. I couldn't believe I was sticking it into someone's chest. I concentrated on maintaining a steep angle of entry, but kept spearing his clavicle instead of slipping beneath it.

"Ow!" he shouted.

"Sorry," I said. S. signaled with a kind of surfing hand gesture to go underneath the clavicle. This time, it went in. I drew back on the syringe. Nothing. She pointed deeper. I went in deeper. Nothing. I withdrew the needle, flushed out some bits of tissue clogging it, and tried again.

"*Ow!*"

Too steep again. I found my way underneath the clavicle once more. I drew the syringe back. Still nothing. He's too obese, I thought. S. slipped on gloves and a gown. "How about I have a look?" she said. I handed her the needle and stepped aside. She plunged the needle in, drew back on the syringe, and, just like that, she was in. "We'll be done shortly," she told the patient.

She let me continue with the next steps, which I bumbled through. I didn't realize how long and floppy the guide wire was until I pulled the coil out of its plastic sleeve, and, putting one end of it into the patient, I very nearly contaminated the other. I forgot about the dilating step until she reminded me. Then, when I put in the dilator, I didn't push quite hard enough, and it was really S. who pushed it all the way in. Finally, we got the line in, flushed it, and sutured it in place.

Outside the room, S. said that I could be less tentative the next time, but that I shouldn't worry too much about how things had gone. "You'll get it," she said. "It just takes practice." I wasn't so sure. The procedure remained wholly mysterious to me. And I could not get over the idea of jabbing a needle into someone's chest so deeply and so blindly. I awaited the X-ray afterward with trepidation. But it came back fine: I had not injured the lung and the line was in the right place.

NOT EVERYONE APPRECIATES the attractions of surgery. When you are a medical student in the operating room for the first time, and you see the surgeon press the scalpel to someone's body and open it like a piece of fruit, you either shudder in horror or gape in awe. I gaped. It was not just the blood and guts that enthralled me. It was also the idea that a person, a mere mortal, would have the confidence to wield that scalpel in the first place.

There is a saying about surgeons: "Sometimes wrong; never in doubt." This is meant as a reproof, but to me it seemed their strength. Every day, surgeons are faced with uncertainties. Information is inadequate; the science is ambiguous; one's knowledge and abilities are never perfect. Even with the simplest operation, it cannot be taken for granted that a patient will come through better off—or even alive. Standing at the operating table, I wondered how the surgeon knew that all the steps would go as planned, that bleeding would be controlled and infection would not set in and organs would not be injured. He didn't, of course. But he cut anyway.

Later, while still a student, I was allowed to make an incision myself. The surgeon drew a six-inch dotted line with a marking pen across an anesthetized patient's abdomen and then, to my surprise, had the nurse hand me the knife. It was still warm from the autoclave. The surgeon had me stretch the skin taut with the thumb and forefinger of my free hand. He told me to make one smooth slice down to the fat. I put the belly of the blade to the skin and cut. The experience was odd and addictive, mixing exhilaration from the calculated violence of the act, anxiety about getting it right, and a righteous faith that it was somehow for the person's good. There was also the slightly nauseating feeling of finding that it took more force than I'd realized. (Skin is thick and springy, and on my first pass I did not go nearly deep enough; I had to cut twice to get through.) The moment made me want to be a surgeon—not an amateur handed the knife for a brief moment but someone with the confidence and ability to proceed as if it were routine.

A resident begins, however, with none of this air of mastery—only an overpowering instinct against doing anything like pressing a knife against flesh or jabbing a needle into someone's chest. On my first day as a surgical resident, I was assigned to the emergency room. Among my first patients was a skinny, dark-haired woman in her late twenties who hobbled in, teeth gritted, with a two-foot-long wooden chair leg somehow nailed to the bottom of her foot. She explained that a kitchen chair had collapsed under her and, as she leaped up to keep from falling, her bare foot had stomped down on a three-inch screw sticking out of one of the chair legs. I tried very hard to look like someone who had not got his medical diploma just the week before. Instead, I was determined to be nonchalant, the kind of guy who had seen this sort of thing a hundred times before. I inspected her foot, and could see that the screw was embedded in the bone at the base of her big toe. There was no bleeding and, as far as I could feel, no fracture.

"Wow, that must hurt," I blurted out, idiotically.

The obvious thing to do was give her a tetanus shot and pull out the screw. I ordered the tetanus shot, but I began to have doubts about pulling out the screw. Suppose she bled? Or suppose I fractured her foot? Or something worse? I excused myself and tracked down Dr. W., the senior surgeon on duty. I found him tending to a car-crash victim. The patient was a mess, and the floor was covered with blood. People were shouting. It was not a good time to ask questions.

I ordered an X-ray. I figured it would buy time and let me check my amateur impression that she didn't have a fracture. Sure enough, getting the X-ray took about an hour, and it showed no fracture—just a common screw embedded, the radiologist said, "in the head of the first metatarsal." I showed the patient the X-ray. "You see, the screw's embedded in the head of the first metatarsal," I said. And the plan? she wanted to know. Ah, yes, the plan.

I went to find Dr. W. He was still busy with the crash victim, but I was able to interrupt to show him the X-ray. He chuckled at the sight of it and asked me what I wanted to do. "Pull the screw out?" I ventured. "Yes," he said, by which he meant "Duh." He made sure I'd given the patient a tetanus shot and then shooed me away.

Back in the examining room, I told her that I would pull the screw out, prepared for her to say something like "You?" Instead she said, "OK, Doctor." At first, I had her sitting on the exam table, dangling her leg off the side. But that didn't look as if it would work. Eventually, I had her lie with her foot jutting off the table end, the board poking out into the air. With every move, her pain increased. I injected a local anesthetic where the screw had gone in and that helped a little. Now I grabbed her foot in one hand, the board in the other, and for a moment I froze. Could I really do this? Who was I to presume?

Finally, I gave her a one-two-three and pulled, gingerly at first and then hard. She groaned. The screw wasn't budging. I twisted, and abruptly it came free. There was no bleeding. I washed the wound out, and she found she could walk. I warned her of the risks of infection and the signs to look for. Her gratitude was immense and flattering, like the lion's for the mouse—and that night I went home elated.

In surgery, as in anything else, skill, judgment, and confidence are learned through experience, haltingly and humiliatingly. Like the tennis player and the oboist and the guy who fixes hard drives, we need practice to get good at what we do. There is one difference in medicine, though: we practice on people.

MY SECOND TRY at placing a central line went no better than the first. The patient was in intensive care, mortally ill, on a ventilator, and needed the line so that powerful cardiac drugs could be delivered directly to her heart. She was also heavily sedated, and for this I was grateful. She'd be oblivious of my fumbling.

My preparation was better this time. I got the towel roll in place and the syringes of heparin on the tray. I checked her lab results, which were fine. I also made a point of draping more widely, so that if I flopped the guide wire around by mistake again, it wouldn't hit anything unsterile.

For all that, the procedure was a bust. I stabbed the needle in too shallow and then too deep. Frustration overcame tentativeness and I tried one angle after another. Nothing worked. Then, for one brief moment, I got a flash of blood in the syringe, indicating that I was in the vein. I anchored the needle with one hand and went to pull the syringe off with the other. But the syringe was jammed on too tightly, so that when I pulled it free I dislodged the needle from the vein. The patient began bleeding into her chest wall. I held pressure the best I could for a solid five minutes, but still her chest turned black and blue around the site. The hematoma made it impossible to put a line through there anymore. I wanted to give up. But she needed a line and the resident supervising me—a second-year this time—was determined that I succeed. After an X-ray showed that I had not injured her lung, he had me try on the other side, with a whole new kit. I missed again, and he took over. It took him several minutes and two or three sticks to find the vein himself and that made me feel better. Maybe she was an unusually tough case.

When I failed with a third patient a few days later, though, the doubts really set in. Again, it was stick, stick, stick, and nothing. I stepped aside. The resident watching me got it on the next try.

Surgeons, as a group, adhere to a curious egalitarianism. They believe in practice, not talent. People often assume that you have to have great hands to become a surgeon, but it's not true. When I interviewed to get into surgery programs, no one made me sew or take a dexterity test or checked to see if my hands were steady. You do not even need all ten fingers to be accepted. To be sure, talent helps. Professors say that every two or three years they'll see someone truly gifted come through a program—someone who picks up complex manual skills unusually quickly, sees tissue planes before others do, anticipates trouble before it happens. Nonetheless, attending surgeons say that what's most important to them is finding people who are conscientious, industrious, and boneheaded enough to keep at practicing this one difficult thing day and night for years on end. As a former residency director put it to me, given a choice between a Ph.D. who had cloned a gene and a sculptor, he'd pick the

Ph.D. every time. Sure, he said, he'd bet on the sculptor's being more physically talented; but he'd bet on the Ph.D.'s being less "flaky." And in the end that matters more. Skill, surgeons believe, can be taught; tenacity cannot. It's an odd approach to recruitment, but it continues all the way up the ranks, even in top surgery departments. They start with minions with no experience in surgery, spend years training them, and then take most of their faculty from these same homegrown ranks.

And it works. There have now been many studies of elite performers—concert violinists, chess grand masters, professional ice-skaters, mathematicians, and so forth—and the biggest difference researchers find between them and lesser performers is the amount of deliberate practice they've accumulated. Indeed, the most important talent may be the talent for practice itself. K. Anders Ericsson, a cognitive psychologist and an expert on performance, notes that the most important role that innate factors play may be in a person's *willingness* to engage in sustained training. He has found, for example, that top performers dislike practicing just as much as others do. (That's why, for example, athletes and musicians usually quit practicing when they retire.) But, more than others, they have the will to keep at it anyway.

I WASN'T SURE I did. What good was it, I wondered, to keep doing central lines when I wasn't coming close to hitting them? If I had a clear idea of what I was doing wrong, then maybe I'd have something to focus on. But I didn't. Everyone, of course, had suggestions. Go in with the bevel of the needle up. No, go in with the bevel down. Put a bend in the middle of the needle. No, curve the needle. For a while, I tried to avoid doing another line. Soon enough, however, a new case arose.

The circumstances were miserable. It was late in the day, and I'd had to work through the previous night. The patient weighed more than three hundred pounds. He couldn't tolerate lying flat because the weight of his chest and abdomen made it hard for him to breathe. Yet he had a badly infected wound, needed intravenous antibiotics, and no one could find veins in his arms for a peripheral IV. I had little hope of succeeding. But a resident does what he is told, and I was told to try the line.

I went to his room. He looked scared and said he didn't think he'd last more than a minute on his back. But he said he understood the situation and was willing to make his best effort. He and I decided that he'd be left sitting propped up in bed until the last possible minute. We'd see how far we got after that.

I went through my preparations: checking his blood counts from the lab, putting out the kit, placing the towel roll, and so on. I swabbed and draped his chest while he was still sitting up. S., the chief resident, was watching me this time, and when everything was ready I had her tip him back, an oxygen mask on his face. His flesh rolled up his chest like a wave. I couldn't find his clavicle with my fingertips to line up the right point of entry. And already he was looking short of breath, his face red. I gave S. a "Do you want to take over?" look. Keep going, she signaled. I made a rough guess about where the right spot was, numbed it with lidocaine, and pushed the big needle in. For a second, I thought it wouldn't be long enough to reach through, but then I felt the tip slip underneath his clavicle. I pushed a little deeper and drew back on the syringe. Unbelievably, it filled with blood. I was in. I concentrated on anchoring the needle firmly in place, not moving it a millimeter as I pulled the syringe off and threaded the guide wire in. The wire fed in smoothly. The patient was struggling hard for air now. We sat him up and let him catch his breath. And then, laying him down one more time, I got the entry dilated and slid the central line in. "Nice job" was all S. said, and then she left.

I still have no idea what I did differently that day. But from then on my lines went in. That's the funny thing about practice. For days and days, you make out only the fragments of what to do. And then one day you've got the thing whole. Conscious learning becomes unconscious knowledge, and you cannot say precisely how.

I HAVE NOW put in more than a hundred central lines. I am by no means infallible. Certainly, I have had my fair share of complications. I punctured a patient's lung, for example—the right lung of a chief of surgery from another hospital, no less—and, given the odds, I'm sure such things will happen again. I still have the occasional case that should go easily but doesn't, no matter what I do. (We have a term for this. "How'd it go?" a colleague asks. "It was a total flog," I reply. I don't have to say anything more.)

But other times everything unfolds effortlessly. You take the needle. You stick the chest. You feel the needle travel—a distinct glide through the fat, a slight catch in the dense muscle, then the subtle pop through the vein wall—and you're in. At such moments, it is more than easy; it is beautiful.

Surgical training is the recapitulation of this process—floundering followed by fragments followed by knowledge and, occasionally, a moment of elegance—over and over again, for ever harder tasks with ever greater risks. At first, you work on the basics: how to glove and gown, how to drape patients, how to hold the knife, how to tie a square knot in a length of silk suture (not to mention how to dictate, work the computers, order drugs). But then the tasks become more daunting: how to cut through skin, handle the electrocautery, open the breast, tie off a bleeder, excise a tumor, close up a wound. At the end of six months, I had done lines, lumpectomies, appendectomies, skin grafts, hernia repairs, and mastectomies. At the end of a year, I was doing limb amputations, hemorrhoidectomies, and laparoscopic gallbladder operations. At the end of two years, I was beginning to do tracheotomies, small-bowel operations, and leg-artery bypasses.

I am in my seventh year of training, of which three years have been spent doing research. Only now has a simple slice through skin begun to seem like the mere start of a case. These days, I'm trying to learn how to fix an abdominal aortic aneurysm, remove a pancreatic cancer, open blocked carotid arteries. I am, I have found, neither gifted nor maladroit. With practice and more practice, I get the hang of it.

Doctors find it hard to talk about this with patients. The moral burden of practicing on people is always with us, but for the most part it is unspoken. Before each operation, I go over to the holding area in my scrubs and introduce myself to the patient. I do it the same way every time. "Hello, I'm Dr. Gawande. I'm one of the surgical residents, and I'll be assisting your surgeon." That is pretty much all I say on the subject. I extend my hand and smile. I ask the patient if everything is going OK so far. We chat. I answer questions. Very occasionally, patients are taken aback. "No resident is doing my surgery," they say. I try to be reassuring. "Not to worry—I just assist," I say. "The attending surgeon is always in charge."

None of this is exactly a lie. The attending *is* in charge, and a resident knows better than to forget that. Consider the operation I did recently to remove a seventy-five-year-old woman's colon cancer. The attending stood across from me from the start. And it was he, not I, who decided where to cut, how to position the opened abdomen, how to isolate the cancer, and how much colon to take.

Yet I'm the one who held the knife. I'm the one who stood on the operator's side of the table, and it was raised to my six-foot-plus height. I was there to help, yes, but I was there to practice, too. This was clear when it came time to reconnect the colon. There are two ways of putting the ends together—handsewing and stapling. Stapling is swifter and easier, but the attending suggested I handsew the ends—not because it was better for the patient but because I had had much less experience doing it. When it's performed correctly, the results are similar, but he needed to watch me like a hawk. My stitching was slow and imprecise. At one point, he caught me putting the stitches too far apart and made me go back and put extras in between so the connection would not leak. At another point, he found I wasn't taking deep enough bites of tissue with the needle to ensure a strong closure. "Turn your wrist more," he told me. "Like this?" I asked. "Uh, sort of," he said.

In medicine, there has long been a conflict between the imperative to give patients the best possible care and the need to provide novices with experience. Residencies attempt to mitigate potential harm through supervision and graduated responsibility. And there is reason to think that patients actually benefit from teaching. Studies commonly find that teaching hospitals have better outcomes than non-teaching hospitals. Residents may be amateurs, but having them around checking on patients, asking questions, and keeping faculty on their toes seems to help. But there is still no avoiding those first few unsteady times a young physician tries to put in a central line, remove a breast cancer, or sew together two segments of colon. No matter how many protections are in place, on average these cases go less well with the novice than with someone experienced.

Doctors have no illusions about this. When an attending physician brings a sick family member in for surgery, people at the hospital think twice about letting trainees participate. Even when the attending insists that they participate as usual, the residents scrubbing in know that it will be far from a teaching case. And if a central line must be put in, a first-timer is certainly not going to do it. Conversely, the ward services and clinics where residents have the most responsibility are populated by the poor, the uninsured, the drunk, and the demented. Residents have few opportunities nowadays to operate independently, without the attending docs scrubbed in, but when we do—as we must before graduating and going out to operate on our own—it is generally with these, the humblest of patients.

And this is the uncomfortable truth about teaching. By traditional ethics and public insistence (not to mention court rulings), a patient's right to the best care possible must trump the objective of training novices. We want perfection without practice. Yet everyone is harmed if no one is trained for the future. So learning is hidden, behind drapes and anesthesia and the elisions of language. And the dilemma doesn't apply just to residents, physicians in training. The process of learning goes on longer than most people know.

I GREW UP in the small Appalachian town of Athens, Ohio, where my parents are both doctors. My mother is a pediatrician and my father is a urologist.

Long ago, my mother chose to practice part time, which she could afford to do because my father's practice became so busy and successful. He has now been at it for more than twenty-five years, and his office is cluttered with the evidence of this. There is an overflowing wall of medical files, gifts from patients displayed everywhere (books, paintings, ceramics with Biblical sayings, hand-painted paperweights, blown glass, carved boxes, a figurine of a boy who, when you pull down his pants, pees on you), and, in an acrylic case behind his oak desk, a few dozen of the thousands of kidney stones he has removed.

Only now, as I get glimpses of the end of my training, have I begun to think hard about my father's success. For most of my residency, I thought of surgery as a more or less fixed body of knowledge and skill which is acquired in training and perfected in practice. There was, I thought, a smooth, upward-sloping arc of proficiency at some rarefied set of tasks (for me, taking out gallbladders, colon cancers, bullets, and appendixes; for him, taking out kidney stones, testicular cancers, and swollen prostates). The arc would peak at, say, ten or fifteen years, plateau for a long time, and perhaps tail off a little in the final five years before retirement. The reality, however, turns out to be far messier. You do get good at certain things, my father tells me, but no sooner do you master something than you find that what you know is outmoded. New technologies and operations emerge to supplant the old, and the learning curve starts all over again. "Three-quarters of what I do today I never learned in residency," he says. On his own, fifty miles from his nearest colleague—let alone a doctor who could tell him anything like "You need to turn your wrist more"—he has had to learn to put in penile prostheses, to perform microsurgery, to reverse vasectomies, to do nerve-sparing prostatectomies, to implant artificial urinary sphincters. He's had to learn to use shock-wave lithotripters, electrohydraulic lithotripters, and laser lithotripters (all instruments for breaking up kidney stones); to deploy Double J ureteral stents and Silicone Figure Four Coil stents and Retro-Inject Multi-Length stents (don't even ask); and to maneuver fiber-optic ureteroscopes. All these technologies and techniques were introduced after he finished training. Some of the procedures built on skills he already had. Many did not.

This is the experience that all surgeons have. The pace of medical innovation has been unceasing, and surgeons have no choice but to give the new thing a try. To fail to adopt new techniques would mean denying patients meaningful medical advances. Yet the perils of the learning curve are inescapable—no less in practice than in residency.

For the established surgeon, inevitably, the opportunities for learning are far less structured than for a resident. When an important new device or procedure comes along, as happens every year, surgeons start by taking a course about it—typically a day or two of lectures by some surgical grandees with a few film clips and step-by-step handouts. You take home a video to watch. Perhaps you pay a visit to observe a colleague perform the operation—my father often goes up to the Cleveland Clinic for this. But there's not much by way of hands-on training. Unlike a resident, a visitor cannot scrub in on cases, and opportunities to practice on animals or cadavers are few and far between. (Britain, being Britain, actually bans surgeons from practicing on animals.) When the pulse-dye laser came out, the manufacturer set up a lab in Columbus where urologists from the area could gain experience. But when my father went there the main experience provided was destroying kidney stones in test tubes filled with a urinelike liquid and trying to penetrate the shell of an egg without hitting the membrane underneath. My surgery department recently bought a robotic surgery device—a staggeringly sophisticated nine-hundred-and-eighty-thousand-dollar robot, with three arms, two wrists, and a camera, all millimeters in diameter, which, controlled from a console, allows a surgeon to do almost any operation with no hand tremor and with only tiny incisions. A team of two surgeons and two nurses flew out to the manufacturer's headquarters, in Mountain View, California, for a full day of training on the machine. And they did get to practice on a pig and on a human cadaver. (The company apparently buys the cadavers from the city of San Francisco.) But even this was hardly thorough training. They learned enough to grasp the principles of using the robot, to start getting a feel for using it, and to understand how to plan an operation. That was about it. Sooner or later, you just have to go home and give the thing a try on someone.

Patients do eventually benefit—often enormously—but the first few patients may not, and may even be harmed. Consider the experience reported by the pediatric cardiac-surgery unit of the renowned Great Ormond Street Hospital, in London, as detailed in the *British Medical Journal* last April. The doctors described their results from three hundred and twenty-five consecutive operations between 1978 and 1998 on babies with a severe heart defect known as transposition of the great arteries. Such children are born with their heart's outflow vessels transposed: the aorta emerges from the right side of the heart instead of the left and the artery to the lungs emerges from the left instead of the right. As a result, blood coming in is pumped right back out to the body instead of first to the lungs, where it can be oxygenated. The babies died blue, fatigued, never knowing what it was to get enough breath. For years, it wasn't technically feasible to switch the vessels to their proper positions. Instead, surgeons did something known as the Senning procedure: they created a passage inside the heart to let blood from the lungs cross backward to the right heart. The Senning procedure allowed children to live into adulthood. The weaker right heart, however, cannot sustain the body's entire blood flow as long as the left. Eventually, these patients' hearts failed, and although most survived to adulthood, few lived to old age.

By the nineteen-eighties, a series of technological advances made it possible to do a switch operation safely, and this became the favored procedure. In 1986, the Great Ormond Street surgeons made the changeover themselves, and their report shows that it was unquestionably an improvement. The annual death rate after a successful switch procedure was less than a quarter that of the Senning, resulting in a life expectancy of sixty-three years instead of forty-seven. But the price of learning to do it was appalling. In their first seventy switch operations, the doctors had a twenty-five-percent surgical death rate, compared with just six percent with the Senning procedure. Eighteen babies died, more than twice the number during the entire Senning era. Only with time did they master it: in their next hundred switch operations, five babies died.

As patients, we want both expertise and progress; we don't want to acknowledge that these are contradictory desires. In the words of one British public report, "There should be no learning curve as far as patient safety is concerned." But this is entirely wishful thinking.

RECENTLY, a group of Harvard Business School researchers who have made a specialty of studying learning curves in industry decided to examine learning curves among surgeons instead of in semiconductor manufacture or airplane construction, or any of the usual fields their colleagues examine. They followed eighteen cardiac surgeons and their teams as they took on the new technique of minimally invasive cardiac surgery. This study, I was surprised to discover, is the first of its kind. Learning is ubiquitous in medicine, and yet no one had ever compared how well different teams actually do it.

The new heart operation—in which new technologies allow a surgeon to operate through a small incision between ribs instead of splitting the chest open down the middle—proved substantially more difficult than the conventional one. Because the incision is too small to admit the usual tubes and clamps for rerouting blood to the heart-bypass machine, surgeons had to learn a trickier method, which involved balloons and catheters placed through groin vessels. And the nurses, anesthesiologists, and perfusionists all had new roles to master. As you'd expect, everyone experienced a substantial learning curve. Whereas a fully proficient team takes three to six hours for such an operation, these teams took on average three times as long for their early cases. The researchers could not track complication rates in detail, but it would be foolish to imagine that they were not affected.

What's more, the researchers found striking disparities in the speed with which different teams learned. All teams came from highly respected institutions with experience in adopting innovations and received the same three-day training session. Yet, in the course of fifty cases, some teams managed to halve their operating time while others improved hardly at all. Practice, it turned out, did not necessarily make perfect. The crucial variable was *how* the surgeons and their teams practiced.

Richard Bohmer, the only physician among the Harvard researchers, made several visits to observe one of the quickest-learning teams and one of the slowest, and he was startled by the contrast. The surgeon on the fast-learning team was actually quite inexperienced compared with the one on the slow-learning team. But he made sure to pick team members with whom he had worked well before and to keep them together through the first fifteen cases before allowing any new members. He had the team go through a dry run before the first case, then deliberately scheduled six operations in the first week, so little would be forgotten in between. He convened the team before each case to discuss it in detail and afterward to debrief. He made sure results were tracked carefully. And Bohmer noticed that the surgeon was not the stereotypical Napoleon with a knife. Unbidden, he told Bohmer, "The surgeon needs to be willing to allow himself to become a partner [with the rest of the team] so he can accept input." At the other hospital, by contrast, the surgeon chose his operating team almost randomly and did not keep it together. In the first seven cases, the team had different members every time, which is to say that it was no team at all. And the surgeon had no pre-briefings, no debriefings, no tracking of ongoing results.

The Harvard Business School study offered some hopeful news. We can do things that have a dramatic effect on our rate of improvement—like being more deliberate about how we train, and about tracking progress, whether with students and residents or with senior surgeons and nurses. But the study's other implications are less reassuring. No matter how accomplished, surgeons trying something new got worse before they got better, and the learning curve proved longer, and was affected by a far more complicated range of factors, than anyone had realized.

This, I suspect, is the reason for the physician's dodge: the "I just assist" rap; the "We have a new procedure for this that you are perfect for" speech; the "You need a central line" without the "I am still learning how to do this." Sometimes we do feel obliged to admit when we're doing something for the first time, but even then we tend to quote the published complication rates of experienced surgeons. Do we ever tell patients that, because we are still new at something, their risks will inevitably be higher, and that they'd likely do better with doctors who are more experienced? Do we ever say that we need them to agree to it anyway? I've never seen it. Given the stakes, who in his right mind would agree to be practiced upon?

Many dispute this presumption. "Look, most people understand what it is to be a doctor," a health policy expert insisted, when I visited him in his office not long ago. "We have to stop lying to our patients. Can people take on choices for societal benefit?" He paused and then answered his question. "Yes," he said firmly.

It would certainly be a graceful and happy solution. We'd ask patients—honestly, openly—and they'd say yes. Hard to imagine, though. I noticed on the expert's desk a picture of his child, born just a few months before, and a completely unfair question popped into my mind. "So did you let the resident deliver?" I asked.

There was silence for a moment. "No," he admitted. "We didn't even allow residents in the room."

ONE REASON I doubt whether we could sustain a system of medical training that depended on people saying "Yes, you can practice on me" is that I myself have said no. When my eldest child, Walker, was eleven days old, he suddenly went into congestive heart failure from what proved to be a severe cardiac defect. His aorta was not transposed, but a long segment of it had failed to grow at all. My wife and I were beside ourselves with fear—his kidneys and liver began failing, too—but he made it to surgery, the repair was a success, and although his recovery was erratic, after two and a half weeks he was ready to come home.

We were by no means in the clear, however. He was born a healthy six pounds plus but now, a month old, he weighed only five, and would need strict monitoring to ensure that he gained weight. He was on two cardiac medications from which he would have to be weaned. And in the longer term, the doctors warned us, his repair would prove inadequate. As Walker grew, his aorta would require either dilation with a balloon or replacement by surgery. They could not say precisely when and how many such procedures would be necessary over the years. A pediatric cardiologist would have to follow him closely and decide.

Walker was about to be discharged, and we had not indicated who that cardiologist would be. In the hospital, he had been cared for by a full team of cardiologists, ranging from fellows in specialty training to attendings who had practiced for decades. The day before we took Walker home, one of the young fellows approached me, offering his card and suggesting a time to bring Walker to see him. Of those on the team, he had put in the most time caring for Walker. He saw Walker when we brought him in inexplicably short of breath, made the diagnosis, got Walker the drugs that stabilized him, coordinated with the surgeons, and came to see us twice a day to answer our questions. Moreover, I knew, this was how fellows always got their patients. Most families don't know the subtle gradations among players, and after a team has saved their child's life they take whatever appointment they're handed.

But I knew the differences. "I'm afraid we're thinking of seeing Dr. Newburger," I said. She was the hospital's associate cardiologist-in-chief, and a published expert on conditions like Walker's. The young physician looked crestfallen. It was nothing against him, I said. She just had more experience, that was all.

"You know, there is always an attending backing me up," he said. I shook my head.

I know this was not fair. My son had an unusual problem. The fellow needed the experience. As a resident, I of all people should have understood this. But I was not torn about the decision. This was my child. Given a choice, I will always choose the best care I can for him. How can anybody be expected to do otherwise? Certainly, the future of medicine should not rely on it.

In a sense, then, the physician's dodge is inevitable. Learning must be stolen, taken as a kind of bodily eminent domain. And it was, during Walker's stay—on many occasions, now that I think back on it. A resident intubated him. A surgical trainee scrubbed in for his operation. The cardiology fellow put in one of his central lines. If I had the option to have someone more experienced, I would have taken it. But this was simply how the system worked—no such choices were offered—and so I went along.

The advantage of this coldhearted machinery is not merely that it gets the learning done. If learning is necessary but causes harm, then above all it ought to apply to everyone alike. Given a choice, people wriggle out, and such choices are not offered equally. They belong to the connected and the knowledgeable, to insiders over outsiders, to the doctor's child but not the truck driver's. If everyone cannot have a choice, maybe it is better if no one can.

IT IS 2 P.M. I am in the intensive-care unit. A nurse tells me Mr. G.'s central line has clotted off. Mr. G. has been in the hospital for more than a month now. He is in his late sixties, from South Boston, emaciated, exhausted, holding on by a thread—or a line, to be precise. He has several holes in his small bowel, and the bilious contents leak out onto his skin through two small reddened openings in the concavity of his abdomen. His only chance is to be fed by vein and wait for these fistulae to heal. He needs a new central line.

I could do it, I suppose. I am the experienced one now. But experience brings a new role: I am expected to teach the procedure instead. "See one, do one, teach one," the saying goes, and it is only half in jest.

There is a junior resident on the service. She has done only one or two lines before. I tell her about Mr. G. I ask her if she is free to do a new line. She misinterprets this as a question. She says she still has patients to see and a case coming up later. Could I do the line? I tell her no. She is unable to hide a grimace. She is burdened, as I was burdened, and perhaps frightened, as I was frightened.

She begins to focus when I make her talk through the steps—a kind of dry run, I figure. She hits nearly all the steps, but forgets about checking the labs and about Mr. G.'s nasty allergy to heparin, which is in the flush for the line. I make sure she registers this, then tell her to get set up and page me.

I am still adjusting to this role. It is painful enough taking responsibility for one's own failures. Being handmaiden to another's is something else entirely. It occurs to me that I could have broken open a kit and had her do an actual dry run. Then again maybe I can't. The kits must cost a couple of hundred dollars each. I'll have to find out for next time.

Half an hour later, I get the page. The patient is draped. The resident is in her gown and gloves. She tells me that she has saline to flush the line with and that his labs are fine.

"Have you got the towel roll?" I ask.

She forgot the towel roll. I roll up a towel and slip it beneath Mr. G.'s back. I ask him if he's all right. He nods. After all he's been through, there is only resignation in his eyes.

The junior resident picks out a spot for the stick. The patient is hauntingly thin. I see every rib and fear that the resident will puncture his lung. She injects the numbing medication. Then she puts the big needle in, and the angle looks all wrong. I motion for her to reposition. This only makes her more uncertain. She pushes in deeper and I know she does not have it. She draws back on the syringe: no blood. She takes out the needle and tries again. And again the angle looks wrong. This time, Mr. G. feels the jab and jerks up in pain. I hold his arm. She gives him more numbing medication. It is all I can do not to take over. But she cannot learn without doing, I tell myself. I decide to let her have one more try.

LIZA MUNDY

A World of Their Own

FROM *THE WASHINGTON POST MAGAZINE*

Would you choose a disability for your child? For some parents, proud members of the vibrant Deaf culture, the birth of a deaf baby is not a cause for despair but a reason to rejoice. Washington Post *reporter Liza Mundy follows a deaf couple awaiting the arrival of a child they are hoping will be happy, healthy, and deaf.*

As her baby begins to emerge after a day of labor, Sharon Duchesneau has a question for the midwife who is attending the birth. Asking it is not the easiest thing, just now. Sharon is deaf, and communicates using American Sign Language, and the combination of intense pain and the position she has sought to ease it—kneeling, resting her weight on her hands—makes signing somewhat hard. Even so, Sharon manages to sign something to Risa Shaw, a hearing friend who is present to interpret for the birth, which is taking place in a softly lit bedroom of Sharon's North Bethesda home.

"Sharon wants to know what color hair you see," Risa says to the midwife.

The midwife cannot tell because the baby is not—quite—visible. He bulges outward during contractions, then recedes when the contraction fades. But now comes another contraction and a scream from Sharon, and the midwife and her assistant call for Sharon to keep pushing but to keep it steady and controlled. They are accustomed to using their voices as a way of guiding women through this last excruciating phase; since Sharon can't hear them, all they can hope is that she doesn't close her eyes.

"Push through the pain!" shouts the midwife.

"Little bit!" shouts her assistant, as Risa frantically signs.

And suddenly the baby is out. One minute the baby wasn't here and now the baby is, hair brown, eyes blue, face gray with waxy vernix, body pulsing with life and vigor. A boy. "Is he okay?" signs Sharon, and the answer, to all appearances, is a resounding yes. There are the toes, the toenails, the fingers, the hands, the eyes, the eyelashes, the exquisite little-old-man's face, contorted in classic newborn outrage. The midwife lays the baby on Sharon and he bleats and hiccups and nuzzles her skin, the instinct to breast-feed strong.

"Did he cry?" signs Sharon, and the women say no, he cried remarkably little.

"His face looks smushed," Sharon signs, regarding him tenderly.

"It'll straighten out," says the midwife.

Presently the midwife takes the baby and performs the Apgar, the standard test of a newborn's condition, from which he emerges with an impressive score of nine out of a possible 10. "He's very calm," she notes as she weighs him (6 pounds 5 ounces), then lays him out to measure head and chest and length. She bicycles his legs to check the flexibility of his hips; examines his testicles to make sure they are descended; feels his vertebrae for gaps.

All in all, she pronounces the baby splendid. "Look how strong he is!" she says, pulling him gently up from the bed by his arms. Which means that it is, finally, possible to relax and savor his arrival. Everyone takes turns holding him: Sharon; her longtime partner, Candace McCullough, who is also deaf, and will be the boy's adoptive mother; their good friend Jan DeLap, also deaf; Risa Shaw and another hearing friend, Juniper Sussman. Candy and Sharon's five-year-old daughter, Jehanne, is brought in to admire him, but she is fast asleep and comically refuses to awaken, even when laid on the bed and prodded. Amid the oohing and aahing someone puts a cap on the baby; somebody else swaddles him in a blanket; somebody else brings a plate of turkey and stuffing for Sharon, who hasn't eaten on a day that's dedicated to feasting. Conceived by artificial insemination 38 weeks ago, this boy, Gauvin Hughes McCullough, has arrived two weeks ahead of schedule, on Thanksgiving Day.

"A turkey baby," signs Sharon, who is lying back against a bank of pillows, her dark thick hair spread against the light gray pillowcases.

"A turkey baster baby," jokes Candy, lying next to her.

"A perfect baby," says the midwife.

"A perfect baby," says the midwife's assistant.

But there is perfect and there is perfect. There is no way to know, yet, whether Gauvin Hughes McCullough is perfect in the specific way that Sharon and Candy would like him to be. Until he is old enough, two or three months from now, for a sophisticated audiology test, the women cannot be sure whether Gauvin is—as they hope—deaf.

SEVERAL MONTHS BEFORE his birth, Sharon and Candy—both stylish and independent women in their mid-thirties, both college graduates, both holders of graduate degrees from Gallaudet University, both professionals in the mental health field—sat in their kitchen trying to envision life if their son turned out not to be deaf. It was something they had a hard time getting their minds around. When they were looking for a donor to inseminate Sharon, one thing they knew was that they wanted a deaf donor. So they contacted a local sperm bank and asked whether the bank would provide one. The sperm bank said no; congenital deafness is precisely the sort of condition that, in the world of commercial reproductive technology, gets a would-be donor eliminated.

So Sharon and Candy asked a deaf friend to be the donor, and he agreed.

Though they have gone to all this trouble, Candy and Sharon take issue with the suggestion that they are "trying" to have a deaf baby. To put it this way, they worry, implies that they will not love their son if he can hear. And, they insist, they will. As Sharon puts it: "A hearing baby would be a blessing. A deaf baby would be a special blessing."

As Candy puts it: "I would say that we wanted to increase our chances of having a baby who is deaf."

It may seem a shocking undertaking: two parents trying to screen in a quality, deafness, at a time when many parents are using genetic testing to screen out as many disorders as science will permit. Down's syndrome, cystic fibrosis, early-onset Alzheimer's—every day, it seems, there's news of yet another disorder that can be detected before birth and eliminated by abortion, manipulation of the embryo or, in the case of in vitro fertilization, destruction of an embryo. Though most deafness cannot be identified or treated in this way, it seems safe to say that when or if it can, many parents would seek to eliminate a disability that affects one out of 1,000 Americans.

As for actively trying to build a deaf baby: "I think all of us recognize that deaf children can have perfectly wonderful lives," says R. Alta Charo, a professor of law and bioethics at the University of Wisconsin. "The question is whether the parents have violated the sacred duty of parenthood, which is to maximize to some reasonable degree the advantages available to their children. I'm loath to say it, but I think it's a shame to set limits on a child's potential."

In the deaf community, however, the arrival of a deaf baby has never evoked the feelings that it does among the hearing. To be sure, there are many deaf parents who feel their children will have an easier life if they are born hearing. "I know that my parents were disappointed that I was deaf, along with my brother, and I know I felt, just for a fleeting second, bad that my children were deaf," says Nancy Rarus, a staff member at the National Association of the Deaf. Emphasizing that she is speaking personally and not on behalf of the association, she adds, "I'm a social animal, and it's very difficult for me to talk to my neighbors. I wish I could walk up to somebody and ask for information. I've had a lot of arguments in the deaf community about that. People talk about 'The sky's the limit,' but being deaf prevents you from getting there. You don't have as many choices."

"I can't understand," she says, "why anybody would want to bring a disabled child into the world."

Then again, Rarus points out, "there are many, many deaf people who specifically want deaf kids." This is true particularly now, particularly in Washington, home to Gallaudet, the world's only liberal arts university for the deaf, and the lively deaf intelligentsia it has nurtured. Since the 1980s, many members of the deaf community have been galvanized by the idea that deafness is not a medical disability, but a cultural identity. They call themselves Deaf, with a capital D, a community whose defining and unifying quality is American Sign Language (ASL), a fluent, sophisticated language that enables deaf people to communicate fully, essentially liberating them—when they are among signers—from one of the most disabling aspects of being deaf. Sharon and Candy share the fundamental view of this Deaf camp; they see deafness as an identity, not a medical affliction that needs to be fixed. Their effort—to have a baby who belongs to what they see as their minority group—is a natural outcome of the pride and self-acceptance the Deaf movement has brought to so many. It also would seem to put them at odds with the direction of reproductive technology in general, striving as it does for a more perfect normalcy.

But the interesting thing is—if one accepts their worldview, that a deaf baby could be desirable to some parents—Sharon and Candy are squarely part of a broader trend in artificial reproduction. Because, at the same time that many would-be parents are screening out qualities they don't want, many are also selecting for qualities they do want. And in many cases, the aim is to produce not so much a superior baby as a specific baby. A white baby. A black baby. A boy. A girl. Or a baby that's been even more minutely imagined. Would-be parents can go on many fertility clinic Web sites and type in preferences for a sperm donor's weight, height, eye color, race, ancestry, complexion, hair color, even hair texture.

"In most cases," says Sean Tipton, spokesman for the American Society of Reproductive Medicine, "what the couples are interested in is someone who physically looks like them." In this sense Candy and Sharon are like many parents, hoping for a child who will be in their own image.

And yet, while deafness may be a culture, in this country it is also an official disability, recognized under the Americans with Disabilities Act. What about the obligation of parents to see that their child has a better life than they did?

Then again, what does a better life mean? Does it mean choosing a hearing donor so your baby, unlike you, might grow up hearing?

Does it mean giving birth to a deaf child, and raising it in a better environment than the one you experienced?

What if you believe you can be a better parent to a deaf child than to a hearing one?

"IT WOULD BE NICE to have a deaf child who is the same as us. I think that would be a wonderful experience. You know, if we can have that chance, why not take it?"

This is Sharon, seven months pregnant, dressed in black pants and a stretchy black shirt, sitting at their kitchen table on a sunny fall afternoon, Candy beside her. Jehanne, their daughter, who is also deaf, and was conceived with the same donor they've used this time, is at school. The family has been doing a lot of nesting in anticipation of the baby's arrival. The kitchen has been renovated, the backyard landscaped. Soon the women plan to rig a system in which the lights in the house will blink one rhythm if the TTY—the telephonic device that deaf people type into—is ringing; another rhythm when the front doorbell rings; another for the side door. They already have a light in the bedroom that will go on when the baby cries.

In one way, it's hard for Sharon and Candy to articulate why they want to increase their chances of having a deaf child. Because they don't view deafness as a disability, they don't see themselves as bringing a disabled child into the world. Rather, they see themselves as bringing a different sort of normal child into the world. Why not bring a deaf child into the world? What, exactly, is the problem? In their minds, they are no different from parents who try to have a girl. After all, girls can be discriminated against. Same with deaf people. Sharon and Candy have faced obstacles, but they've survived. More than that, they've prevailed to become productive, self-supporting professionals. "Some people look at it like, 'Oh my gosh, you shouldn't have a child who has a disability,'" signs Candy. "But, you know, black people have harder lives. Why shouldn't parents be able to go ahead and pick a black donor if that's what they want? They should have that option. They can feel related to that culture, bonded with that culture."

The words "bond" and "culture" say a lot; in effect, Sharon and Candy are a little like immigrant parents who, with a huge and dominant and somewhat alien culture just outside their door, want to ensure that their children will share their heritage, their culture, their life experience. If they are deaf and have a hearing child, that child will move in a world where the women cannot fully follow. For this reason they believe they can be better parents to a deaf child, if being a better parent means being better able to talk to your child, understand your child's emotions, guide your child's development, pay attention to your child's friendships. "If we have a hearing child and he visits a hearing friend, we'll be like, 'Who is the family?'" says Candy. "In the deaf community, if you don't know a family, you ask around. You get references. But with hearing families, we would have no idea."

They understand that hearing people may find this hard to accept. It would be odd, they agree, if a hearing parent preferred to have a deaf child. And if they themselves—valuing sight—were to have a blind child, well then, Candy acknowledges, they would probably try to have it fixed, if they could, like hearing parents who attempt to restore their child's hearing with cochlear implants. "I want to be the same as my child," says Candy. "I want the baby to enjoy what we enjoy."

Which is not to say that they aren't open to a hearing child. A hearing child would make life rich and interesting. It's just hard, before the fact, to know what it would be like. "He'd be the only hearing member of the family," Sharon points out, laughing. "Other than the cats."

"DID YOU WEIGH yourself?"

"What?"

"Did you weigh yourself?"

"Yes," says Sharon. It's a few weeks before the baby's birth, and Sharon has taken the Metro to Alexandria for a prenatal checkup. Wearing a long black skirt and loose maroon blouse, she has checked in at the BirthCare and Women's Health Center and has been ushered into an examining room, where she now shifts, bulky, in her seat.

"How are you feeling?" the midwife asks.

"Tired today," says Sharon. Often, Sharon brings her hearing friend Risa Shaw to interpret at checkups, but today she's relying on her own ability to speak and read lips. Reading lips is something Sharon does remarkably well. She developed the skill on her own. Growing up, she was also enrolled in speech therapy, where a progression of therapists fitted her with hearing aids, shouted into her ear, sent her home to practice talking in front of a mirror because her "a" was too nasal, and generally instilled in her, she says now, a sense of constant failure. On one level, the therapy worked: When she speaks, she does so with fluency and precision.

But even the following small exchange shows what an inexact science lip-reading is. "This is our first visit?" the midwife says, looking at her chart.

"What?" Sharon replies, peering to follow the movement of her tongue and teeth and lips.

"This . . . is . . . my . . . first . . . visit . . . with . . . you," says the midwife, speaking more slowly.

"Oh," says Sharon, who has seen other midwives on previous visits. "Yes."

"Let's see—we are at 36 weeks, huh? So today we need to do an internal exam and also do the culture for beta strep. You're having a home birth, right? So do you have the oxygen?" "What?"

"The oxygen?"

"What?"

The midwife gestures to indicate an oxygen tank, one of the supplies they need to have on hand at home.

"No."

This gives some sense of what life has been like for Sharon, who was raised in what's known as the oralist tradition. Which is to say, she was raised to function in the hearing world as best she could, without exposure to sign language or to other deaf people, except her mother. Like her mother, Sharon was born with some residual hearing but experienced hearing loss to the point where, at eight or so, she was severely deaf. Her father, Thomas, a professor of economics at the University of Maine, can hear, and so can her younger sister, Anne. In this family Sharon was referred to as "hearing impaired" or "hard of hearing," rather than "deaf." She attended public school in Bangor; there was a special classroom for deaf kids, and Sharon stayed as far away from it as possible.

"I find it very hard to say now," says Sharon. "Sometimes my speech therapist would want me to meet the other deaf children, and it was an embarrassment. I didn't want to be identified with them. I didn't want my friends to look at me as if I was different."

Those friendships were relatively easy when she was young, riding bikes and running around, but became much harder in adolescence, where so much of friendship is conducted verbally, in groups, which are impossible to lip-read. She got by. "I played field hockey, I did layout for the yearbook, it looked like I did fine, but inside I always felt there was something wrong with me. I remember someone would ask what kind of music I liked, and I didn't know what the cool answer would be. I used to make my sister write down the words to the most popular songs."

She grew up feeling that her sister was normal and that she was flawed, a feeling, she says, exacerbated by her father, who pushed her to speak. She knows he meant well, and Sharon functioned so ably, it's easy to see why his expectations for her were high. But those standards filled her with a desire to meet them and a chronic sense of falling short. "Once when I was 11 or 12, my family went to a restaurant to eat, and I wanted to have milk to drink, and I was trying to tell the waitress and she couldn't understand me. I think I tried maybe two or three times, and she kept looking at me like I was speaking Chinese. I looked at my father like: 'Help me out here.' And he was like: 'Go ahead. Say it again.' "

Another time, she says, her father told her that if she ever had children, she should check with a geneticist to assess the risk that her baby, like her, would be deaf. "I felt put down, like it would be bad if my child was deaf, or it was a negative thing to bring a deaf child into the world," she says. "I took it personally."

And high school, compared with what came later, was easy. Having done well academically, Sharon enrolled at the University of Virginia. She tries to convey the numbing isolation of that experience; of being at a huge college full of strangers; being from out of state; being deaf; straining to catch names; feeling at sea in dorms or at parties; sitting at the front of big classes, tape-recording the lecture and then taking the tape to a special office to be typed, then returning, alone, to her room with a 30- or 40-page transcript. For a hearing person, perhaps the best analogy would be to imagine yourself in a foreign country where you understand the language only slightly; where comprehension will not get better no matter how hard you try. "I got," she says, "very tired of that."

She gravitated to a major in medical ethics, and in that department she met a professor who urged her to learn sign and meet some deaf people. Sharon resisted; he persisted, pointing out that if she learned sign, she could interview deaf people as part of her research. So she relented, went to Gallaudet for a summer of sign lessons, and realized that her professor's argument had been a ploy. "The first day I got there, I knew that it wasn't about taking it for school. It was for myself," she says. She returned to U-Va., graduated, got an internship in the bioethics department at the National Institutes of Health. But her heart and mind were in continuing her sign lessons and becoming part of the deaf community. The writer Oliver Sacks, in his book about deafness, *Seeing Voices,* has described American Sign Language, for deaf people encountering it for the first time, as coming home.

"It was the best time," she says. "There were so many wonderful things about it. About deaf people, about signing. People understood me. I didn't have to explain myself. I didn't have to fake it. It was a positive thing to be deaf at Gallaudet.

"That summer," she continues, beginning to weep, "really changed my life, my hopes and my dreams and my future. It changed everything.

"Before that," she says, "I couldn't think about the future. I felt so lost."

Some of this lostness had to do with her sexual identity. She had never dated men much, and at Gallaudet she became increasingly aware of herself as a lesbian. A fellow student recognized this, took her out to some bars, helped her come out. She went on to pursue a master's in the Gallaudet counseling department; it was during that period that she met Candy, a slender, vivacious woman with a taste for leather jackets and hip, flared trousers. At the time, Candy drove a Honda Prelude with a sound system that had—deaf people experience music through vibrations—really hot woofers.

Unlike Sharon, Candy had been brought up signing, the child of deaf parents, but that doesn't mean her upbringing was easier. Neither of her parents finished high school. Her father was a printer, the classic deaf profession; historically, to be deaf often meant to be relegated to industrial work—factories and print shops being among the few places where it is an advantage not to hear. They lived in northern California, where for a while she was put in a special deaf classroom in an inner-city Oakland school, where signing was not permitted in class. Candy was so bright she worked through the entire third-grade math textbook in a weekend, but she felt the expectations of her were very low (some kids with deafness are also born with other disorders, so the range of abilities in a deaf classroom is very broad). She transferred to a special school for deaf kids, but—finding that easy, too—transferred again to a hearing high school, where she attended classes with an interpreter. But an interpreter can't help a high schooler make friends. No teenage conversation can survive the intrusion of third-party interpretation, and Candy, unlike Sharon, was not able to speak for herself. Profoundly deaf from birth, she had no residual hearing to help her figure out how a voice should sound. Even with speech therapy, she'd learned early on that hearing people could not understand her when she spoke. "So," she says now, "I stopped talking."

At lunch the interpreter would take a break, and Candy, unable to talk to anyone, would go to the library and do her homework. On weekends, she studied or worked at the library shelving books. "I was the perfect student," she says, so from high school she went to the University of California at Berkeley. Like Sharon, she found college grindingly lonely. Her first year she met Ella Mae Lentz, a deaf poet who composes in ASL. Lentz suggested Candy transfer to Gallaudet. Like Sharon, Candy felt a deaf school would be academically inferior. But, Lentz pointed out, a crucial part of college is having friends. Candy had already come out as a lesbian; her mother was upset, so it occurred to Candy that 3,000 miles away might be a good place to be. So she transferred, and like Sharon, she has never looked back. The women, who have been together for nearly 10 years, moved in with each other, then bought a house with their close friend Jan DeLap. At some point Sharon spoke of a dream she'd once had but dismissed: to have children. She assumed they couldn't, not because they were deaf but because they were lesbians. It is not Candy's nature to dismiss dreams. " 'Can't' isn't in my vocabulary," she says. So they found a donor, a friend of Candy's who comes from five generations of deafness. In Sharon's family there are four generations on her mother's side. Once she was pregnant, a genetic counselor predicted that based on these family histories, there was a 50–50 chance her child would be deaf. Heads for a deaf child, tails for hearing.

The very first time—with Jehanne—the coin came up heads.

Candy usually signs with both hands, using facial expressions as well as signs. This is all part of ASL, a physical language that encompasses the whole body, from fingers to arms to eyebrows, and is noisy, too: There is lots of clapping and slapping in ASL, and in a really great conversation, it's always possible to knock your own eyeglasses off.

When she drives, though, Candy also signs one-handed, keeping the other hand on the wheel. Chatting with Sharon, she maneuvers her Volvo through Bethesda traffic and onto I-270, making her way north toward Frederick, home to the Maryland School for the Deaf. State residential schools have played a huge role in the development of America's deaf community. Historically, deaf children often left their homes as young as five and grew up in dorms with other deaf kids. This sometimes isolated them from their families but helped to create an intense sense of fellowship among the deaf population, a group that, though geographically spread out, is essentially a tribe, a small town, a family itself.

Now that people are more mobile, families with deaf children often relocate near a residential school for the deaf, where the young children are more likely to be day students. Jehanne is one; today she's waiting for them in a low corridor inside the elementary school building at MSD, petite, elfin, dimpled, with tousled brown hair and light brown, almost amber eyes. Essentially, the baby Sharon is carrying represents a second effort that they're making because the first was so successful. (Candy tried to have their second child, but a year of efforts didn't take.) At her own infant audiology test, Jehanne was diagnosed as profoundly deaf. In their baby book, under the section marked "first hearing test," Candy wrote, happily, "Oct. 11, 1996—no response at 95 decibels—DEAF!"

This afternoon, Jehanne greets her mothers and begins immediately to sign. She has been signed to since birth and, unlike her mothers, has been educated from the start in sign. At five she is beginning to read English quite well; when they're riding in the car, she'll notice funny shop names, like Food Lion and For Eyes. But she is also fluent in ASL, more fluent even than Sharon.

The women have arrived to visit Jehanne's kindergarten classroom, which in most ways is similar to that of any other Maryland public school; the kids are using flashcards to learn about opposites, conducting experiments to explore concepts like wet and dry, light and heavy. The classes are small, and teachers are mostly deaf, which is something new; years ago, even at MSD, deaf people weren't permitted to teach the young kids, because it was believed that sign would interfere with their learning to read. Now that's all changed. Sign is used to teach them reading. They learn science in sign; they sign while doing puzzles, or gluing and pasting, or coloring, or working in the computer lab.

There is a speech therapy class, but it's optional, and a far cry from the ones that Sharon and Candy remember, where laborious hours were spent blowing on feathers to see the difference between a "b" and a "p." In general, Sharon and Candy have tried not to make what they see as the mistakes their own parents did. Sharon, for example, resents having been made to wear hearing aids and denied the opportunity to learn sign, while Candy—who really wanted to try a hearing aid when she was little—was told by her father that she couldn't because it would be expensive and pointless, anyway. Trying to chart a middle course, they let Jehanne decide for herself whether she wanted to try a hearing aid; she did, one summer when attending camp at Gallaudet. It was hot pink. She wore it about a week.

Similarly, they left it up to her whether to take speech therapy; since she is much more profoundly deaf than Sharon, it is unlikely that she will ever have speech as clear as Sharon's. But she wanted to take the class; when they asked why, she told them that it was fun. Now they understand why. When Jehanne and another friend are pulled out for speech class on this day, they make their way down the hall to a classroom where the children enact a mock Thanksgiving dinner. The teacher passes out plastic turkey and mashed potatoes and bread; as they pretend to eat, enjoying the role-playing, the teacher signs and speaks.

"Now we're going to do what with our napkins?" she says as the two girls look up at her. "Put it in our l-l-l-l-l-ap." She exaggerates the sound, so they can see how an "l" is made. The girls learn speech by watching her and then trying to imitate the tongue and lip movements they see. At such a young age, the sounds that emerge are vague and tentative.

"Now we need a knife," she says, and Jehanne makes a sound like "nuh."

"Knife."

"Nuh."

"Would you like some water?"

Jehanne makes a good-faith effort to say "yes, please," pursing her lips and wiggling her tongue to come out with a "pl."

Candy and Sharon watch intently, concerned not about Jehanne's speech but about the teacher's style of signing. At one point she tells Jehanne to lay out her napkin, but because the sign isn't the classic ASL sign, Jehanne looks at her blankly. "Oh well," says Sharon later. "It's good for her to know that not everybody is a fluent signer." They inspect the computer lab, chatting with the school Webmaster, whom they know; he and his wife are the parents of one of Jehanne's classmates. For Sharon and Candy, one of the great advantages of having a deaf child is that it gives them a built-in social life. Like most parents, they socialize a lot with the parents of their children's friends, and at MSD, many of the parents are deaf. They also see the school as one way to ensure that Jehanne doesn't endure the loneliness and isolation that they did. By raising her among deaf children, they feel she's getting a much stronger start in life.

And they are every bit as ambitious for Jehanne as any parent would be for a child. Afterward, the women talk to the principal, who is also deaf. They tell her they are happy with the school, with a few caveats: They wish she had a little more self-directed time; they wish the weekly written reports were more detailed. Jehanne, who is clearly an outstanding student, is also just a tiny bit klutzy, no big deal, but even so they'd like to hear some details from the gym teacher. Her last report, for gym, was checked "needs improvement." "Needs improvement? What does that tell me?" signs Candy. "We've taken her to dance class, soccer; we swim each week, she does yoga! What more do you want us to do?" Laughing, Sharon and Candy talk about the fact that Jehanne is one of those kids who haven't figured out how to swing; she's still trying to get the pumping motion. It's an interesting moment. To most parents, hearing would seem a much more important ability, in the grand scheme of things, than pumping. But that's not how Candy and Sharon see it.

"She's a sweetheart," says the principal soothingly. "She's a role model. She's in with such a nice group of friends." The principal has known most of these kids almost since the day they were born. At MSD, deaf infants qualify for a weekly morning class. When they are two, they go to preschool. Their education—with small classrooms, extra teachers, transportation—is free, paid for with public funds.

So advantageous is MSD, in fact, that one of the things Candy and Sharon think about is how much more a hearing child would cost. If the baby is hearing, they'll have to pay for day care. For preschool. Even, if they find they don't agree with the teaching philosophy of the public schools, for private school. "It's awful to think that, but it'll be more expensive!" Sharon acknowledges.

But—while deaf children do receive some financial advantages—they point out that deaf children give back, in ways that are complex and impossible to predict. Take Candy and Sharon themselves: Both work at home as counselors, seeing deaf clients and, often, hearing family members. Not only do they provide the deaf with clear, accessible mental health care; Sharon also finds that hearing patients sometimes open up more for a therapist who is not herself "perfect." And hearing parents of deaf children are often "relieved to come and see a deaf therapist," Sharon finds. "They're like, 'Oh, you went to college! Oh, that means my children can do that!' They're afraid the child will be on the street selling pencils."

So sure, Jehanne's education may cost the public more. But deaf children, Sharon argues, make a society more diverse, and diversity makes a society more humane. Plenty of individuals and groups receive public support, and if you start saying which costs are legitimate and which aren't, well, they believe, it's a slippery slope.

"Do you think this baby's hearing?" Candy asks Sharon afterward, when they are having lunch in downtown Frederick.

"I don't know," says Sharon. "I can say that I hope the baby's deaf, but to say I feel it's deaf, no."

They are talking about an old saying in the deaf community: If the mother walks into a place with loud music, and the baby moves, the baby is hearing. "If you base it on that, I do think it's deaf," says Sharon.

"I just say to myself that the baby's deaf," Candy says. "I talk as if the baby's deaf. If the baby's hearing, I'll be shocked."

"You better be prepared," Sharon tells her. "With Jehanne, I prepared myself. It could happen." Thinking about it, she speculates: "A hearing child would force us to get out and find out what's out there for hearing children. Maybe that would be nice."

Candy looks at her, amazed.

"It's not that it's my preference," says Sharon. "But I'm trying to think of something positive."

EXACTLY TWO WEEKS after his birth, Gauvin (pronounced Go-VAHN, as in French) is sleeping in a Moses basket, luminous and pink and tiny. He continues to sleep, undisturbed, when Jan DeLap turns on the disposal and Candy loudly grates cheese with the salad shooter. But when Sharon begins to set the table, opening cupboards and clattering plates, he shifts, clenches his fists and stretches. Jehanne pretends to test his hearing, making a noise like "buh-buh-buh," and he writhes a little. When she is relaxed and around people she loves—as now—Jehanne makes noises all the time, a low, constant, happy humming.

The more relaxed a deaf household is, the noisier it is. Around hearing people, deaf people are careful to control the sounds they make, but when they're alone they can let go. When Sharon wants Candy, she calls her by stomping the floor. When the cats get on the table, Jan lets out a hair-raising whoop. It doesn't always work. One of the cats, they believe, is hard of hearing. The veterinarian disagrees. "He thought we were projecting," Sharon says.

Dinner tonight is burritos. Gauvin, who is turning out to be a very easy baby, is still sleeping, so they can eat uninterrupted and chat with Jehanne. In school, Jehanne's class is reading *The Very Busy Spider,* which involves animals saying "baaa" and "neigh" and "meow," sounds that none of the kids has heard. And so today, Jehanne tells them, they learned about animal sounds.

"What does a duck say?" asks Candy.

"Oink, oink," signs Jehanne.

"No!" signs Candy, amused. "Quack! Quack!"

"What does a rooster say?" she asks. Jehanne is stumped, and so, for a minute, is everybody else.

"Oh yeah!" somebody remembers. "Cock-a-doodle-doo!"

After dinner, it's story time. The house is full of books. Downstairs are shelf after shelf of novels, nonfiction and clinical textbooks, even a shelf dedicated to the English language, everything from dictionaries of English usage to the *Pocket Dictionary of American Slang*. They are constantly buying books for Jehanne; tonight they're reading *Elizabite: Adventures of a Carnivorous Plant* and *Blueberries for Sal*.

Candy is tonight's designated reader. She signs the stories in ASL, sometimes with both hands, sometimes with one and using the other to point to the words. Candy is such a beautiful, vivid signer that the stories seem to possess her, and she them. Hands fluttering, face mobile and focused on Jehanne, Candy is Little Sal's mother putting berries in her tin pail, plink plank plunk; she is Mother Bear, separated from her cub; she is both of the babies, Little Sal and Little Bear, looking for their moms. Jehanne watches, rapt; Jan watches, rapt; Sharon, who is now breast-feeding Gauvin on a couch in the living room, watches, rapt. A deep contentment falls over the household. "And the bear went over and she heard the rumbling of Little Bear in the bushes, and she knew that it was her baby, and they went down the mountain, eating berries and storing them up for winter!" Candy finishes.

After Jehanne goes to bed, they take out an inking kit to record Gauvin's footprint in his baby book. Like most second babies, Gauvin doesn't have the extensive archives that his older sibling does. His baby book is still somewhat sparse, whereas Jehanne's is crammed full of tiny writing. Under "baby's first words," Candy noted that at about 11 months—the time most babies would say their first word—Jehanne signed "fan." Soon came "swing," and "more," and "light." In the section where the parents are to write their aspirations for the baby, Candy wrote: "Jehanne can plan her own future. Seeing her happy is all that is important to us."

It is an open question, however, to what extent Jehanne can plan her own future. Candy and Sharon say that it will be okay with them if she goes to Gallaudet, but okay, too, if she wants to go to a hearing college. Though it would be harder for her to participate, say, in student government or athletics or dorm life, they think otherwise she would manage. And after that? The opportunities, they believe, are unlimited. Recently, though, Jehanne and Sharon were talking about astronauts, and Jehanne asked whether a deaf person can be an astronaut. Sharon was obliged to tell her no. Astronauts, she explained, need to communicate by radio. "That's not nice!" Jehanne said. "It's not nice that deaf people can't be astronauts!" Sharon told her maybe someday astronauts will be able to use video.

But with the exception of that—and, probably, of the classic childhood ambition, president—they do feel that Jehanne can be what she wants. She has electronic communications to help her; e-mail has made a huge difference to deaf people. She'll have what they feel is the solid foundation of an education anchored by sign. They think she'll have what they never had: strong self-esteem, a powerful belief in herself. She'll have the considerable legal protection of the Americans with Disabilities Act, which forbids employment discrimination.

Not that the ADA can solve everything. Candy, who is in the final stages of getting her doctorate in psychology, needs to do a yearlong internship at a hospital or other workplace. She plans to counsel both deaf and hearing patients; plans, in short, to be a psychologist like any other. This means two things. It means an interpreter will need to be hired. It also means she is competing mostly with hearing applicants. When she sends off her résumé, there is no indication she is deaf; at Gallaudet, most of the students in her graduate program are hearing people who plan to work with the deaf. But if she gets an interview, she has to e-mail the prospective employer, to discuss her need for an interpreter.

"If I go and they aren't interested," she says, "how do I know why? It's hard sometimes to know whether discrimination is taking place, or not."

"Some deaf people think it's a hard life," reflects Candy, whose grandfather wanted to be a pilot but was prevented by deafness. "But some people think the world is open."

"Did you ever want to be a policeman?" she asks Jan, whose father was a cop. Jan, who is 60, had a deaf mother but a hearing father, so she grew up around hearing relatives, and from them was exposed to music. When she was seven, she saw a movie about an opera singer. "I told my friends that I wanted to be an opera singer," Jan recalls. "My cousin was like, 'You can't be an opera singer. You're deaf!' I think that at that point I thought, 'I'm deaf now but maybe I can be hearing later.' "

"I remember wanting to be a lawyer," says Candy. "And then my teacher said that a deaf person can't do it. And later it wasn't my area of interest."

Now, Jan mentions, there are quite a few deaf lawyers. They have a friend who is one. In the courtroom she makes use of something called real-time captioning. There are technical advances every day. But technology doesn't help a deaf person who is standing next to a hearing person who can't sign. It will never completely bridge what is, still, an enormous gap. Jehanne has a neighbor she plays with, a hearing girl she's known almost since birth. The mothers agree that as they get older, it's getting harder and harder for the girls to communicate, and they get together less and less.

"What I wonder," Jan says at one point, "is whether they'll eliminate the deaf gene. Maybe they'll be able to pluck out the deaf gene. Maybe there will be no more deaf people." They sit contemplating this. It isn't out of the question. Members of another disabled group were taken by surprise when the gene that causes their condition was discovered: Now, a child with achondroplasia, or disproportionately short arms and legs—also known as dwarfism—can be identified in utero. And, if the parents don't want a child with dwarfism, the fetus can be aborted. The community of "little people," which has its own association, its own Web site, a strong tradition in Hollywood, and a powerful fellowship, has been left contemplating its children destroyed, its numbers dwindling, its existing members consigned to a narrowing life of freakishness and isolation. Such a fate could—it's possible—befall the deaf. The situation illustrates how in this country, at this cultural moment, disabled people are exposed to two powerful but contradictory messages. One says: You are beautiful. You are empowered. The other says: You are deficient. You may be snuffed out.

"Maybe there will be no more deaf children," Jan says.

"Except," says Candy, "for those of us who choose to make more deaf children."

As the weeks go by, Gauvin starts staying awake more. His eyes, blue and wide, start tracking more; he watches his mothers, and Jehanne, with an intensity that they believe is characteristic of deaf children. They sign to him in deaf "motherese"; like a hearing mother speaking in a high-pitched, singsong voice, they sign slowly, with exaggerated gestures. In mid-December they take him to Gallaudet for a show. In the auditorium there are people signing across the room, people signing from the floor to people in the balcony.

In this group Gauvin is admired like a crown prince. Friends, colleagues and former classmates come to peek inside the sling in which Sharon is carrying him, and, inevitably, to inquire whether he is deaf. "How many of you are deaf?" asks the emcee, and Jan—half-joking, half-serious—motions to Sharon to raise Gauvin's hand.

There are many more admirers: In December the sperm donor comes for a visit, as he does about twice a year. Then, after Christmas, Sharon's father, Thomas, arrives. Sharon's mother died of breast cancer not long after Sharon graduated from U-Va., so he is here with the woman who is now his companion, Caroline Dane. Both of them are hearing. Also visiting are Candy's mother, Diana, who is deaf; Sharon's sister, Anne, who is hearing; Anne's boyfriend, Paul, who is hearing. That means there are four hearing people in the house and five deaf people. Plus Gauvin, whatever team he ends up on. Jehanne moves from one group to another, but usually gravitates toward people who are signing, because she has no way, save by gesturing, to communicate with her hearing relatives.

Sharon is the pivot point, the only one who can translate, which is exhausting for her. She has to keep lip-reading and talking and signing, almost simultaneously. When an interpreter arrives to interpret for this article, the entire group—all 10 of them—crowds into the living room and sits, talking intently, for two hours.

It is the first time they have been able to fully express themselves to one another, the first time Sharon has ever had someone to interpret a conversation with her own family. The first time she didn't have to strain to understand what her father said, or her sister. Much of it is funny and fond: It turns out that Thomas, cleaning out his attic, recently found some of the song lyrics that Anne transcribed for Sharon, back when both were girls. "You saved those?" says Sharon. "Why?" Then Anne remembers how she would interpret for Sharon on the phone.

"I remember when that boy asked you to the prom," says Anne, who is six years younger than Sharon, her hair lighter brown, her face illuminated by the same quizzical expressions, the same seriousness, the same faintly Gallic beauty.

"You interpreted that?" Sharon says, laughing.

"Yes!" says Anne, who also remembers that whenever Sharon didn't want to go out with a boy, Anne was the one who had to tell him.

"Do you remember that time we were having an argument, and I called you 'deaf'?" Anne says.

"You weren't happy. A lot has changed."

Together, the sisters try to excavate some of their mother's history, find out why she never signed: Both Sharon and her mother struggled to lip-read each other, mother and daughter divided rather than united by deafness, their common bond. Eventually Sharon confronts her father with what she sees as the central mistake her parents made in her upbringing. "I can look back now," she signs, "and say that things would have been different if I had learned to sign, or been exposed to deaf culture. Growing up, if I got 60 percent of a conversation, I felt like that was good. Some of those behaviors are still with me. In groups of signers, they may be signing really fast and even if I'm not getting it all I'm like, 'This is good enough.' I still don't like asking people to repeat. I'm just used to not getting everything."

Later, sitting with her father, she asks, "Did you feel bad when I said that I wished it had been different when I was growing up?"

"No," says her father, a solid, deliberative man with glasses who has brought Jehanne a University of Maine sweatshirt. "We all think about that. We all feel that way about our parents."

IN TRYING TO KNOW how to think about Sharon and Candy's endeavor, there are any number of opinions a person might have. Any number of abstract ideas a person might work through in, say, an ethics course. Are the women being selfish? Are they inflicting too much hardship on the child? How does one think of them compared with, say, a mother who has multiple embryos implanted in the course of fertility treatments, knowing that this raises the likelihood of multiple births and, with it, birth defects in some or all of the babies? Morally, how much difficulty can a parent impose on a child in order to satisfy the desire to have a child, or to have a certain kind of child?

A person can think about this, and think about it, but eventually will run up against the living, breathing fact of the child herself. How much difficulty have Sharon and Candy imposed on Jehanne? They haven't deafened her. They've given life to her. They've enabled her to exist. If they had used a hearing donor, they would have had a different child. That child would exist, but this one wouldn't. Jehanne can only exist as what she is: Jehanne, bright, funny, loving, loved, deaf.

And now what about Gauvin, who, at three months, already resembles his sister? He has the same elfin face shape, the same deep dimples when he smiles. On his head is a light fuzz of hair; bulkier now, alert and cheery, he's wearing gray overalls and groovy red leather sneakers. The question that will be answered this February afternoon, at Children's National Medical Center, is whether Gauvin, like Jehanne, is deaf. Whether the coin has landed on the same side twice. By now, Gauvin has had an initial hearing screening, which he failed. They considered this good news, but not conclusive. From there he was referred to this one, which is more sophisticated. The preliminaries take awhile. Sharon lays Gauvin in a crib and a technician applies conductive paste at points around his head, then attaches electrodes to the paste. He needs to be asleep for the test, in which microphones will be placed in his ears and a clicking noise sent through the wires. Through the electrodes, a machine will monitor the brain response. If the waves are flat, there is no hearing. He stirs and cries, so Sharon breast-feeds him, wires dangling from his head, until he falls asleep. The technician slips the microphone in his ear, turns on the clicking noise—up and up, louder and louder—and the two women look at the computer screen. Even at 95 decibels, a sound so loud that for hearing people it's literally painful, the line for the left ear is flat. But there is a marked difference in the right. For softer sounds the line is flat, but at 75 decibels there is a distinct wave. The technician goes to fetch the doctor, and the mothers contemplate their sleeping son, who, it appears, might be neither deaf nor hearing but somewhere in between.

The doctor, Ira Weiss, bustles in; he is a white-haired, stocky man, jovial and accustomed to all sorts of parents, hearing and deaf, happy and sobbing.

The technician points to the wave and suggests that perhaps it represents some noise that Gauvin himself was making. "No," says the doctor, "I think it's not just noise." Sharon looks up at Candy and lets out a little breath. The doctor disappears to get a printout of the results, then returns, reading it. Gauvin, he says, "has a profound hearing loss in his left ear and at least a severe hearing loss in his right ear.

"It does appear," he adds, "that his right ear has some residual hearing. There might be some usable hearing at this time. Given the mother's history, it will probably get worse over time. If you want to take advantage of it, you should take advantage of it now. Right now it's an ear that could be aided, to give him a head start on spoken English. Obviously, he's going to be a fluent signer."

At this stage, Weiss says later, a hearing parent would probably try a hearing aid, in the hope that with it, that right ear could hear something. Anything. A word, here and there. A loud vowel. Maybe just enough residual sound to help him lip-read. Maybe just enough to tell him when to turn his head to watch someone's lips. Hearing parents would do anything—anything—to nudge a child into the hearing world. Anything—anything—to make that child like them. For a similar reason, Sharon and Candy make the opposite choice. If he wants a hearing aid later, they'll let him have a hearing aid later. They won't put one on him now. After all, they point out, Sharon's hearing loss as a child occurred at below 40 decibels, which meant that under certain conditions she could make out voices, unaided. Gauvin's, already, is far more severe than hers. Bundling Gauvin up against the cold, they make their way down the corridor, and into the car, and home, where they will tell Jehanne, and Jan, and friends, and family, a sizable group, really, that wants to know. He is not as profoundly deaf as Jehanne, but he is quite deaf. Deaf enough.

FLOYD SKLOOT

The Melody Lingers On

FROM *SOUTHWEST REVIEW*

As researchers try to get a better fix on its causes and pathology, Alzheimer's disease remains a terrifying mystery. With candor and tenderness, the poet and writer Floyd Skloot observes the toll the disease is taking on his ninety-one-year-old mother, who literally and figuratively may have forgotten the words but can remember the tune.

A t ninety-one, deep in dementia, my mother no longer remembers her life. Thoughts drift as though in zero gravity, bumping occasionally against a few stray bits of memory, but nothing coheres. Her two husbands, her late son, all the cousins and community acquaintances who filled her days, her ambitions and achievements, her travels and yearnings—almost everything has floated away from her grasp, mere debris.

"Was I happily married?" she asked last week, when my wife, Beverly, and I took her out for coffee and snacks. Before I could answer, she added, "*Oh how we somthinged on the hmmm hmm we were wed.* Dear, was I ever on the stage?"

I nodded and said, "On the radio too."

"I was on the radio?" She smiled, closed her eyes and sang, "*Birds gotta swim, fish gotta fly, da-dada-da one man da-da die.*" Then she lifted a fragment of blueberry muffin and said, "Was I ever on the stage?"

It's not just her distant past that's gone. What happened two minutes ago is as lost as what happened during the twenty-seven years she lived in Manhattan, the twenty years she lived in Brooklyn or the forty-four years she lived on Long Island. Now that she's in Oregon, she doesn't know she ever lived elsewhere. Sometimes she believes her Portland nursing home is a beachfront hotel, just as she sometimes believes I am her late brother.

What's become apparent, though, is that she still knows songs. She retains many lyrics, snatches that may get confused but are easily recognizable, and when the lyrics are missing the melodies remain. She loves to sing, sings on key and with zest, and I can't help wondering why song has hung on so tenaciously while her life memories have not.

It's tempting to take the psychological approach: She never was very happy with her life, but she was happy dreaming of stardom as a torch singer. She was happy knowing she'd had a brief career singing on radio in the mid-1930s, where her five-minute program on WBNX in the Bronx aired opposite Rudy Valee. In the chemical bath of her mind, she always transformed a few years of apprentice costume work in the legitimate theater, and an assortment of roles in local community theater, into a protracted career in the *Thee-a-ter*. No question: she loved performing. I remember how extravagantly she accompanied herself on the piano, sliding along the bench to reach her notes, stomping the pedals, rising and sitting again, going through her brief repertoire before erupting with gusto at the end as a signal for applause. According to this psychological approach, my mother forgets what she needs to forget, and is left with song.

But such an explanation isn't really convincing, not when the evidence of deep organic brain damage is so apparent in her activities of daily living. She cannot dress herself, needs reminding during a meal if she is to continue eating, cannot process new information. Her failures of memory are not choices, not driven by subconscious needs. It must be that, unlike personal memories or the recall of facts, such things as song lyrics are stored in a part of her brain that has, so far, escaped the ravages of her dementia.

AS A RESULT of advances in neuroscience, the pattern of my mother's losses can be pinpointed biologically. First, there's the sheer diminishment in her overall cognitive capacity. In his book *Searching for Memory*, Harvard psychologist Daniel L. Schacter says that "overall brain mass steadily shrinks as we enter our sixties and seventies, at roughly 5 percent to 10 percent per decade." So my mother's brain has probably lost about a quarter of its size by now. In addition, "blood flow and uptake of oxygen both decrease significantly" and there is "widespread loss of neurons in the cortex," a major site of memory storage. In

Alzheimer's patients, the shrinking brain also becomes clotted with plaques and tangles, and there is further neuron loss in the hippocampus, a part of the brain associated with the ability to remember the ongoing incidents in our lives. This set of compounding pathologies explains most of her symptoms, but not the curious endurance of those songs. It's most likely that my mother's lifelong joy in performing, and the powerful emotional forces associated with it for her, have enabled the deeper storage of lyric and melody in her amygdala. This almond-shaped organ in the inner brain is critical for forming and sustaining emotional memories. Though most often spoken of in connection with persistent, enduring traumatic memories, it also is responsible for enduring positive memories. This is where our most vivid memories reside, etched there by a mix of chemical and physical processes that ensure their endurance. I suspect that my mother's amygdala has not yet been overtaken by her disease process. This would explain not only the persistence of her song repertoire, but the relative calmness and sweetness she still manifests. As David Shenk notes in his book *The Forgetting*, when the "amygdala becomes compromised, control over primitive emotions like fear, anger, and craving is disrupted; hostile emotions and bursts of anxiety may occur all out of proportion to events, or even out of nowhere."

My mother is not there yet. In trying to reduce her symptoms to these objective clinical explanations, I know I'm trying to cushion myself from the changes she's undergone and from what lies ahead. But this is my mother, not some interesting case history in a neurology text. This is the woman who fought to allow my birth, eight years after my older brother's, overcoming my father's continuing resistance. The woman who recited nonsense verse to me, sounds I still remember fifty years later though she does not, though she can no longer always remember who I am: *Nicky nicky tembo, whatso rembo, wudda wudda boosky, hippo pendro, national pom pom.* The woman, so miserable and disappointed throughout her life, filled with anger, volatile, friendless in old age, who now in dementia has grown sweet and accommodating, happy to greet the day, who has come back to song.

Those songs of hers, which routinely interrupt any effort at communication, are in fact signs of hope. They represent an enduring part of her past, connected with the rare joy in her life, which is why they linger when so much is gone. I must learn to welcome rather than be annoyed by them. In many ways, they're all we have left of her.

———

LIKE THE MASS of her brain, the physical structure of my mother's body is also shrinking. At her tallest, about 5′ 1″, the top of her head used to be level with the middle of my forehead; now she comes up to my throat. She was always wide, too, a solid and blocky woman whose flowing outfits didn't disguise her figure as she'd hoped. She took up room despite being small. But now she has lost both water and mass. Her once swollen legs have slimmed; she sags and looks frail. It's as though my mother is pulling herself in around a diminishing core, the dwindling autobiographical self she's losing touch with, and closing down before my eyes.

She was moving slowly toward our table in the coffee shop, inching her walker along, taking a few steps and stopping. When she reached the table where Beverly was placing napkins and spoons, she looked around with a smile, let go of her side rails, tilted her head heavenward and sang, "*S'wonderful, s'marvelous, la da da.*"

Her voice still comes from down near her chest, the way it's supposed to, a richly resonating smoky contralto. It's almost as deep as my own off-key tenor. But as a young singer, my mother was a soprano. There are three surviving 78 rpm records from her radio show that prove it. She was called "The Melody Girl of the Air" on a program hosted by an old family friend, and once a week she sang a few standards for him. George Gershwin was alive then and Gershwin was her favorite. There were times when she hinted at a romance with him, never going quite so far as to say they'd dated, but implying that a certain dashing young composer—whom she was not at liberty to name but who had a dowdy lyricist brother—was once very interested in her.

A solid fifty years of unfiltered Chesterfields transformed my mother's voice and, though she stopped smoking in her early seventies, those cigarettes remain audible now in her gravelly tones. But she can and does still belt out the tunes, holding nothing back. This dynamic and deeper voice is how I remember her singing. I never could make sense out of those old records, the high pitched girlishness, the piercing delivery. In my hearing, she sang dark and windy.

There was always a well-tuned mahogany piano against a living room wall in our various apartments. Its lid was shut, its music deck empty, its surfaces without dust or fingerprints. No one was allowed to sit on its bench or open the keyboard lid, much less touch the keys or press the pedals down. No examining the sheet music hidden inside the bench. She wasn't sure she wanted us even to *look* at the piano.

By the 1950s, as I was growing up, my mother's performance repertoire had been condensed to five tunes that she would play in the same order. She seldom sang more than one refrain and chorus, took no requests, brooked no singing along. She would consent to entertain at the end of small dinner parties or holiday meals, perching on the bench and holding her chin up until there was total silence. Then she struck a chord *fortissimo* and launched herself into performance. First came the Gershwin portion of the program, "They Can't Take That Away from Me," " 'S Wonderful" and "Our Love Is Here to Stay." Then she did Rodgers and Hart's "Bewitched, Bothered, and Bewildered" from *Pal Joey* and finished with her signature song, Jerome Kern and Oscar Hammerstein's "Can't Help Lovin' Dat Man," from *Show Boat.* No encores. She was still, it seemed, tied to the fifteen-minute radio show format.

I see now that her songs were songs of love, joyful love. Along with fame as a performer, this was the other great unfulfilled yearning of her life. It was not there with the man she married first, who died in 1961, the man she spoke of in my hearing as "your father the butcher." Nor was it there apparently with her second husband, a kind and gentle widower, the man she spoke of as "that nice, handsome fellow." After his death, when she moved into a retirement hotel overlooking the boardwalk, my mother had a succession of boyfriends but none without glaring faults—too old and bent, too devoted to children and grandchildren, too working-class, too senile. Now, from within her own dementia, one of the main themes woven through her rambling speech is love, joyful love. Was she happily married? Does she have a boyfriend? Are Beverly and I married? Are we happy? Is the nurse married? The young man behind the Starbucks counter? Can we help my mother find a new boyfriend?

Even as a child, I sensed something that made me very uncomfortable with my mother's recitals. It wasn't just the showy way she played, or the too-familiar spontaneity of her moves. It had to do with the look on her face, a rapturous hunger, and the sudden exposure of her deepest, most obsessive wishes. She leaped off her piano bench like a nearly drowned diver suddenly bursting to the surface, head back, mouth wide open, and I imagine her longing was palpable to everyone. There was something brazenly sad about her selection of romantic hits, a sadness I failed to appreciate for most of my life. She must have wanted what she could never have, what few people ever have, and she hadn't let go of the need: idealized romantic love. Her playlist was a litany of failed dreams.

Those failed dreams and her overall sense of disgruntlement seem to have shrunk now too. With the fading of memory and life story has come an apparent narrowing of mood. From the outside, seeing how she is now, this phenomenon suggests a compensation for her shattering losses, and I hope that's how it works for her. I know it could have been otherwise. Like so many people with Alzheimer's or deep dementia, she could long ago have become even angrier and more tormented, hostile and restless.

MY MOTHER LOOKED DOWN at the coffee in her cup, unsure what to do with it. She gazed into its tawny surface and blinked. Only when she looked away did she, as though triggered by signals from a more instinctive zone of her brain, lift the cup toward her lips. I helped her steady it.

"Was I ever on the stage, dear?" she asked again.

I responded automatically: "I whistle a happy tune . . . ," and she beamed, picking up the tune itself, humming along, nodding firmly. One way to look at this, I've realized, is to consider song lyrics as my mother's native tongue. Tonal and melodious, its beauties of sound offset by the banality of its linguistic content. Well, beauties of sound when she sings it, not when I do.

"What comes next?" she asked.

"I hold my head . . . ," and she nods again, taking over, finishing out the sentence after her own fashion: "*so no one da dee da I forget.*"

From the mid-1950s to the mid-1970s, my mother was active in community theater. She was usually cast in small singing roles. I remember her playing Melba Snyder in *Pal Joey*, doing the striptease number "Zip" in the basement of our Brooklyn synagogue. She played the nanny, Gooch, in *Auntie Mame*, where she was dressed up by Mame and her friend Vera for one night as a swinger (singing "I lived! I lived! I lived!"). She was King Mongkut's first wife, Lady Thiang, in *The King and I,* shunted aside for younger wives, the romance of marriage gone though she still admires her husband and tells Anna so ("Something Wonderful"). Taking on a non-singing role, she was Yente the Matchmaker in *Fiddler on the Roof.* What these roles all have in common is their tangential relationship to passion: a cynical stripper, an unglamorous nanny spruced up for a quick taste of the sexy high life, a queen spurned and settling for grandeur instead of romance, an old woman whose business is brewing love for others. Ironically, even as she got to fulfill her desire for performing, the roles she played re-enacted romantic failure and disappointment.

I performed with her on occasion, when no reasonable excuse could be found. When she was in *The King and I,* I was ten and played one of the king's children, learning my one schoolroom speech ("What is that green over there?"), rehearsing the March of the Siamese Children, singing my brief solo

("Suddenly I'm bright and breezy") in "Getting to Know You." At thirteen, during a horrifying cabaret-style local fund-raiser, I sang a duet with her, the dutiful Sonny Boy climbing upon my mother's knee though we were the same size.

When she wasn't part of a play's cast, she still became engaged in the productions. She attended rehearsals to play the piano or read cues or kibbitz. She painted sets. Resorting to her earliest contributions to the theater, she helped design costumes.

I remember her working on hat designs for a production of *Guys and Dolls.* She would glue buttons onto blank greeting cards, paint black dots for eyes in the buttonholes and red dots for mouths, add a few ink strokes for hair. Then she snipped bits of fabric and feathers to resemble hats, pasted them onto the crowns of the buttons and made tiny adjustments with toothpicks. Below the buttons, she drew the shape of necks, then added scarves or ties. It was possible for her to devote four or five intense hours a night to this work, cigarettes smoldering in her abalone shell ashtray. The finished illustrations would be spread out over a card table to dry or for further modification. Finally, she would bring them to rehearsal, stacked in a shoe box, and get herself ready for another round.

My mother was, clearly, a trouper. I cannot remember her being as focused or as sprightly as she was at her design work or within the acting company. She saved all the reviews from our local paper, all the programs, and most of the scripts. I found them in a storage locker when she moved into the retirement hotel and, just glancing at them, felt myself swamped with the scents and sounds of her theatrical life.

AS WE TURNED onto Boundary Avenue, bringing my mother back to the nursing home after our outing, Beverly said, "This is the street where you live." Then, as though on cue, all three of us started singing Lerner and Loewe's "On the Street Where You Live": *I have often walked down the street before.* My mother's voice fractured into laughter and she could hardly keep singing. Besides, she didn't exactly have the words anymore. So she went into scat: *Doo doo doo doo do, la da dee doo da, knowing I'm doo be la doo wee oh.* We pulled up to the front door, all three of us cackling at our mutual cleverness.

I've noticed during recent visits that my mother's repertoire has actually expanded. She's no longer limited to the Big Five Hits. Now she'll bring out songs I never heard her sing before, like "Fly Me to the Moon" or "Anything Goes," which I recognize from the sustained melodies more than from the snippet of lyrics she can muster. Dixon and Henderson's "Bye, Bye Blackbird" from 1926, Gus Kahn's "Makin' Whoopee" from 1928. She sings Yiddish songs, too, all new to me, songs she must have learned during her childhood, when Yiddish was spoken at home and in the Upper West Side neighborhoods where she lived. I haven't heard her speak a word of Yiddish since we moved from Brooklyn in 1957 and cannot remember her ever singing in the language that might have marked her as marginal. She also now has the melodies for some Hebrew tunes she must be picking up during Sabbath services at the nursing home. I don't believe they come from her memory tune-bank because she never went to the synagogue except for social or theatrical events, and I haven't heard her utter a word in Hebrew before.

Much as I'm amazed to hear her dredge up songs from her childhood or youth, it's the phenomenon of new songs—"Adon Olam," for instance, and "Hatikvah"—that astounds me. Perhaps this means that, because she still connects so powerfully to music, she can somehow learn and remember fresh material, at least song material, particularly melodies, though in conversation she cannot remember the question she asked a moment before, or whether we told her what state we live in, or if we're married. Asked if she has been to Sabbath services, she says, "No, they don't have them here." But they do, and she has, and the melodies have stuck.

She also comes up with songs I know she's heard in my lifetime but I hadn't realized she remembered. And she delivers them with genuine glee. *Be down to la da in a taxi baby, doo dah be-dee dee in your hay dee hay.* Gradually, I've been discovering that this is an opportunity for conversation of a sort. While it's not possible for me to ask her questions and get meaningful answers, or share information with her about the life Beverly and I are leading, or even go over memories of childhood with her, we can approximate the give and take of conversation through song. "What are those?" I'll say, pointing to the necklace of beads she's made during a crafts session. "Baubles? Bangles?" And she'll say, "*Bubbles, Bangles bright shiny beads la da dee dah.*" Or I'll hum the opening notes from "If I Were a Rich Man" and she will pick up the song from there.

I'm beginning to find a solace in this exchange. We have the rhythm of conversation, if not the content. A form of give and take that enables us still to feel connected by words, or at least by meaningful sounds. "The song is ended," as her favorite songwriter wrote, "but the melody lingers on." We are holding on to the melody of contact. And they can't take that away from me, from us, at least not yet.

FRANK WILCZEK

The World's Numerical Recipe

FROM *DAEDALUS*

> *The phrase "music of the spheres" has passed into metaphor, but it was originally coined to describe celestial motion. The physicist Frank Wilczek wonders whether the music of the spheres is to be found not in the grand movements of the heavens but in the tiny workings of the atom.*

Twentieth-century physics began around 600 B.C. when Pythagoras of Samos proclaimed an awesome vision.

By studying the notes sounded by plucked strings, Pythagoras discovered that the human perception of harmony is connected to numerical ratios. He examined strings made of the same material, having the same thickness, and under the same tension, but of different lengths. Under these conditions, he found that the notes sound harmonious precisely when the ratio of the lengths of string can be expressed in small whole numbers. For example, the length ratio 2:1 sounds a musical octave, 3:2 a musical fifth, and 4:3 a musical fourth.

The vision inspired by this discovery is summed up in the maxim "All Things Are Number." This became the credo of the Pythagorean Brotherhood, a mixed-sex society that combined elements of an archaic religious cult and a modern scientific academy.

The Brotherhood was responsible for many fine discoveries, all of which it attributed to Pythagoras. Perhaps the most celebrated and profound is the

Pythagorean Theorem. This theorem remains a staple of introductory geometry courses. It is also the point of departure for the Riemann-Einstein theories of curved space and gravity.

Unfortunately, this very theorem undermined the Brotherhood's credo. Using the Pythagorean Theorem, it is not hard to prove that the ratio of the hypotenuse of an isosceles right triangle to either of its two shorter sides cannot be expressed in whole numbers. A member of the Brotherhood who revealed this dreadful secret drowned shortly afterward, in suspicious circumstances. Today, when we say $\sqrt{2}$ is irrational, our language still reflects these ancient anxieties.

Still, the Pythagorean vision, broadly understood—and stripped of cultic, if not entirely of mystical, trappings—remained for centuries a touchstone for pioneers of mathematical science. Those working within this tradition did not insist on whole numbers, but continued to postulate that the deep structure of the physical world could be captured in purely conceptual constructions. Considerations of symmetry and abstract geometry were allowed to supplement simple numerics.

In the work of the German astronomer Johannes Kepler (1570-1630), this program reached a remarkable apotheosis—only to unravel completely.

Students today still learn about Kepler's three laws of planetary motion. But before formulating these celebrated laws, this great speculative thinker had announced another law—we can call it Kepler's zeroth law—of which we hear much less, for the very good reason that it is entirely wrong. Yet it was his discovery of the zeroth law that fired Kepler's enthusiasm for planetary astronomy, in particular for the Copernican system, and launched his extraordinary career. Kepler's zeroth law concerns the relative size of the orbits of different planets. To formulate it, we must imagine that the planets are carried about on concentric spheres around the Sun. His law states that the successive planetary spheres are of such proportions that they can be inscribed within and circumscribed about the five Platonic solids. These five remarkable solids—tetrahedron, cube, octahedron, dodecahedron, icosahedron—have faces that are congruent equilateral polygons. The Pythagoreans studied them, Plato employed them in the speculative cosmology of the *Timaeus*, and Euclid climaxed his *Elements* with the first known proof that only five such regular polyhedra exist.

Kepler was enraptured by his discovery. He imagined that the spheres emitted music as they rotated, and he even speculated on the tunes. (This is the source of the phrase "music of the spheres.") It was a beautiful realization of the Pythagorean ideal. Purely conceptual, yet sensually appealing, the zeroth law seemed a production worthy of a mathematically sophisticated Creator.

To his great credit as an honest man and—though the concept is anachronistic—as a scientist, Kepler did not wallow in mystic rapture, but actively strove to see whether his law accurately matched reality. He discovered that it does not. In wrestling with the precise observations of Tycho Brahe, Kepler was forced to give up circular in favor of elliptical orbits. He couldn't salvage the ideas that first inspired him.

After this, the Pythagorean vision went into a long, deep eclipse. In Newton's classical synthesis of motion and gravitation, there is no sense in which structure is governed by numerical or conceptual constructs. All is dynamics. Newton's laws inform us, given the positions, velocities, and masses of a system of gravitating bodies at one time, how they will move in the future. They do not fix a unique size or structure for the solar system. Indeed, recent discoveries of planetary systems around distant stars have revealed quite different patterns. The great developments of nineteenth-century physics, epitomized in Maxwell's equations of electrodynamics, brought many new phenomena within the scope of physics, but they did not alter this situation essentially. There is nothing in the equations of classical physics that can fix a definite scale of size, whether for planetary systems, atoms, or anything else. The world-system of classical physics is divided between initial conditions that can be assigned arbitrarily, and dynamical equations. In those equations, neither whole numbers nor any other purely conceptual elements play a distinguished role.

Quantum mechanics changed everything.

Emblematic of the new physics, and decisive historically, was Niels Bohr's atomic model of 1913. Though it applies in a vastly different domain, Bohr's model of the hydrogen atom bears an uncanny resemblance to Kepler's system of planetary spheres. The binding force is electrical rather than gravitational, the players are electrons orbiting around protons rather than planets orbiting the Sun, and the size is a factor 10^{-22} smaller; but the leitmotif of Bohr's model is unmistakably "Things Are Number."

Through Bohr's model, Kepler's idea that the orbits that occur in nature are precisely those that embody a conceptual ideal emerged from its embers, reborn like a phoenix, after three hundred years' quiescence. If anything, Bohr's model conforms more closely to the Pythagorean ideal than Kepler's, since its preferred orbits are defined by whole numbers rather than geometric constructions. Einstein responded with great empathy and enthusiasm, referring to Bohr's work as "the highest form of musicality in the sphere of thought."

Later work by Heisenberg and Schrödinger, which defined modern quantum mechanics, superseded Bohr's model. This account of subatomic matter is less tangible than Bohr's, but ultimately much richer. In the Heisenberg-

Schrödinger theory, electrons are no longer particles moving in space, elements of reality that at a given time are "just there and not anywhere else." Rather, they define oscillatory, space-filling wave patterns always "here, there, and everywhere." Electron waves are attracted to a positively charged nucleus and can form localized standing wave patterns around it. The mathematics describing the vibratory patterns that define the states of atoms in quantum mechanics is identical to that which describes the resonance of musical instruments. The stable states of atoms correspond to pure tones. I think it's fair to say that the musicality Einstein praised in Bohr's model is, if anything, heightened in its progeny (though Einstein himself, notoriously, withheld his approval from the new quantum mechanics).

The big difference between nature's instruments and those of human construction is that her designs depend not on craftsmanship refined by experience, but rather on the ruthlessly precise application of simple rules. Now if you browse through a textbook on atomic quantum mechanics, or look at atomic vibration patterns using modern visualization tools, "simple" might not be the word that leaps to mind. But it has a precise, objective meaning in this context. A theory is simpler the fewer nonconceptual elements, which must be taken from observation, enter into its construction. In this sense, Kepler's zeroth law provided a simpler (as it turns out, too simple) theory of the solar system than Newton's, because in Newton's theory the relative sizes of planetary orbits must be taken from observation, whereas in Kepler's they are determined conceptually.

From this perspective, modern atomic theory is extraordinarily simple. The Schrödinger equation, which governs electrons in atoms, contains just two nonconceptual quantities. These are the mass of the electron and the so-called fine-structure constant, denoted α, that specifies the overall strength of the electromagnetic interaction. By solving this one equation, finding the vibrations it supports, we make a concept-world that reproduces a tremendous wealth of real-world data, notably the accurately measured spectral lines of atoms that encode their inner structure. The marvelous theory of electrons and their interactions with light is called quantum electrodynamics, or QED.

In the initial modeling of atoms, the focus was on their accessible, outlying parts, the electron clouds. The nuclei of atoms, which contain most of their mass and all of their positive charge, were treated as so many tiny (but very heavy!) black boxes, buried in the core. There was no theory for the values of nuclear masses or their other properties; these were simply taken from experiment.

That pragmatic approach was extremely fruitful and to this day provides the working basis for practical applications of physics in chemistry, materials science, and biology. But it failed to provide a theory that was in our sense simple, and so it left the ultimate ambitions of a Pythagorean physics unfulfilled.

Starting in the early 1930s, with electrons under control, the frontier of fundamental physics moved inward, to the nuclei. This is not the occasion to recount the complex history of the heroic constructions and ingenious deductions that at last, after fifty years of strenuous international effort, fully exposed the secrets of this inaccessible domain. Fortunately, the answer is easier to describe, and it advances and consummates our theme.

The theory that governs atomic nuclei is quantum chromodynamics, or QCD. As its name hints, QCD is firmly based on quantum mechanics. Its mathematical basis is a direct generalization of QED, incorporating a more intricate structure supporting enhanced symmetry. Metaphorically, QCD stands to QED as an icosahedron stands to a triangle. The basic players in QCD are quarks and gluons. For constructing an accurate model of ordinary matter just two kinds of quarks, called up and down or simply u and d, need to be considered. (There are four other kinds, at least, but they are highly unstable and not important for ordinary matter.) Protons, neutrons, π mesons, and a vast zoo of very short-lived particles called resonances are constructed from these building blocks. The particles and resonances observed in the real word match the resonant wave patterns of quarks and gluons in the concept-world of QCD, much as states of atoms match the resonant wave patterns of electrons. You can predict their masses and properties directly by solving the equations.

A peculiar feature of QCD, and a major reason why it was hard to discover, is that the quarks and gluons are never found in isolation, but always in complex associations. QCD actually predicts this "confinement" property, but that's not easy to prove.

Considering how much it accounts for, QCD is an amazingly simple theory, in our objective sense. Its equations contain just three nonconceptual ingredients: the masses of the u and d quarks and the strong coupling constant α_s, analogous to the fine structure constant of QED, which specifies how powerfully quarks couple to gluons. The gluons are automatically massless.

Actually even three is an overestimate. The quark-gluon coupling varies with distance, so we can trade it in for a unit of distance. In other words, mutant QCDs with different values of α_s generate concept-worlds that behave identically, but use different-sized metersticks. Also, the masses of the u and d quarks turn out not to be very important, quantitatively. Most of the mass of strongly interacting particles is due to the pure energy of the moving quarks and gluons they contain, according to the converse of Einstein's equation,

$m = E/c^2$. The masses of the u and d quarks are much smaller than the masses of the protons and other particles that contain them.

Putting all this together, we arrive at a most remarkable conclusion. To the extent that we are willing to use the proton itself as a meterstick, and ignore the small corrections due to the u and d quark masses, QCD becomes a theory with *no nonconceptual elements whatsoever*.

Let me summarize. Starting with precisely four numerical ingredients, which must be taken from experiment, QED and QCD cook up a concept-world of mathematical objects whose behavior matches, with remarkable accuracy, the behavior of real-world matter. These objects are vibratory wave patterns. Stable elements of reality—protons, atomic nuclei, atoms—correspond, not just metaphorically but with mathematical precision, to pure tones. Kepler would be pleased.

This tale continues in several directions. Given two more ingredients, Newton's constant G_N and Fermi's constant G_F, which parametrize the strength of gravity and of the weak interaction, respectively, we can expand our concept-world beyond ordinary matter to describe virtually all of astrophysics. There is a brilliant series of ideas involving unified field theories and supersymmetry that might allow us to get by with just five ingredients. (Once you're down to so few, each further reduction marks an epoch.) These ideas will be tested decisively in coming years, especially as the Large Hadron Collider (LHC) at CERN, near Geneva, swings into operation around 2007.

On the other hand, if we attempt to do justice to the properties of many exotic, short-lived particles discovered at high-energy accelerators, things get much more complicated and unsatisfactory. We have to add pinches of many new ingredients to our recipe, until it may seem that rather than deriving a wealth of insight from a small investment of facts, we are doing just the opposite. That's the state of our knowledge of fundamental physics today—simultaneously triumphant, exciting, and a mess.

The last word I leave to Einstein:

> I would like to state a theorem which at present can not be based upon anything more than upon a faith in the simplicity, i.e., intelligibility, of nature: there are no *arbitrary* constants . . . that is to say, nature is so constituted that it is possible logically to lay down such strongly determined laws that within these laws only rationally completely determined constants occur (not constants, therefore, whose numerical value could be changed without destroying the theory).

MARCELO GLEISER

Emergent Realities in the Cosmos

If we are the universe's sole intelligent species, asks the physicist and astronomer Marcelo Gleiser, then what must we do to be good citizens of the cosmos?

There is a creative tension in the cosmos. We feel it every time we look at Nature, and we feel it within ourselves. It is revealed in the smallest of details, a dewdrop balancing on the tip of a leaf on an early fall morning, the hexagonal symmetry of snowflakes, resulting from water's molecular structure and heat dissipation. And it is revealed in large-scale natural phenomena, a lightning strike ripping across the sky during a stormy night, or in stars burning their entrails in order to survive the inexorable crush of their own gravity. Our collective history can be told as an effort to represent and make sense of this creative tension, this constant dance of chaos and order that shapes the world.

We have created countless stories, drawings, dances, and rituals in search of meaning, in search of answers. We look at the cosmos with a mixed sense of awe and wonder, of terror and devotion. And we want to know. How can something come from nothing? What is the origin of all things? Can order emerge by itself, without a guiding hand? Is beauty a mere accident of Nature, or is there a deeper meaning to it? Why do we crave beauty, as junkies a drug? What is it that makes us plant gardens, compose poems and symphonies, create mathematical theorems and equations? Why can't we be content simply by eating, procreating, and sleeping? These are questions that bridge and expand our ways of knowing, being part of cutting-edge scientific research, philosophical meditation, religious prayer, and artistic output. We have an unquenchable urge to understand who we are and what is our place in this vast Universe. In many ways, it is through this search for answers that we define ourselves. By asking, by wanting to know, we define what it means to be human. And, although the answers may vary, just as cultures vary from place to place and time to time, many questions are the same, and remain, to a large extent, unanswered.

Modern science has developed a comprehensive narrative describing the emergence of material structures in the Universe. Although many of the details and fundamental questions remain open, we now can claim with certitude that the history of the cosmos traces an increasing complexification of its living and nonliving inhabitants, of the hierarchical development of form and function from the simple to the complex. Thus, at very early times, when the Universe was extremely hot and dense, matter was in the form of its most basic constituents, the indivisible elementary particles. As the Universe expanded and cooled, attractive forces between the different particles made clustering possible: protons and neutrons emerged from binding quarks, atomic nuclei from binding protons and neutrons, light atoms from binding atomic nuclei and electrons, galaxies from huge collapsing hydrogen clouds, stars from smaller hydrogen-rich clouds within these galaxies until, eventually, living beings emerged in at least one of the billions of solar systems spread across the cosmos.

The scientific account describing the emergence of complex material structures has enjoyed enormous success. Cosmology is now a data-driven branch of physics, as opposed to even two decades ago. However, in spite of this success, or perhaps because of it, several fundamental questions have surfaced that defy present knowledge. Among the most fascinating of these questions are questions of origins: the origin of the cosmos, the origin of life, and the origin of the mind. The answers to these questions, even if presently unknown, are all related to the issue of emergence: How is it that structures self-organize to the point of generating extremely sophisticated complex behavior? Be it a surging cosmos out of a primordial soup of cosmoids, a simple living being made of millions of organic macromolecules, or a thinking being, capable of wondering about his or her own origins and of pondering moral dilemmas, the emergence of complexity encompasses some of the most awesome and least understood natural phenomena.

These three origin questions may be compressed into a single one: "How come us?" This is the kind of exasperating question that makes most scientists throw in the towel. A common answer is "Who cares?" After all, there may not be a reason at all; we may be here simply as the result of a random sequence of accidents, the right-size planet, with the right amount of water, at the right distance from a moderate-size star, and so on. "The Universe may be full of Earth-like planets with other forms of intelligent life," the argument proceeds. Indeed, it is quite possible that the Universe is filled with Earth-like planets, some of them with similar amounts of water and Earth-like atmospheric compositions. Possibly, several will also have some form of living beings. If Earth is a demonstrative example, life is very resilient and can adapt to very adverse circumstances. But intelligent life is a whole other story. (By intelligent I mean a species capable of self-reflection and with the ability for abstract thinking.)

Evolutionary arguments claiming that natural selection necessarily leads to intelligence are flawed. Consider the history of life in the only place we actually know it, Earth. The dinosaurs were here for about 150 million years and showed no signs of decline or of intelligence. Intelligence may be a sufficient condition for dominating the food chain, but it is not a necessary one. It took a devastating collision with a 10-kilometer-wide asteroid 65 million years ago to decimate the dinosaurs, together with 40 percent of all life-forms on Earth. Ironically, the mammals, which up to that point were pretty much insignificant, survived and flourished in the wake of this cataclysm. In a very real sense, we are here due to this catastrophic collision.

Life is an experiment in emergent complexity: we may know what the ingredients are, but we cannot predict its detailed outcome (and we still cannot repeat it in the laboratory). Intelligent life is certainly a very rare outcome. This goes against everything we have learned over the last 400 years of modern science, that the more we know about the Universe the less unique we seem to be. True, we live in one amongst billions of other galaxies in the visible Universe, each of them with billions of stars. True, the matter that makes up people and stars is subdominant; most of the matter that permeates the cosmos is not made of protons and electrons, but of something else that does not shine, as matter making up stars does. Our location in the cosmos and our material composition are not of great cosmic relevance or special. But our minds are. As far as we know, there aren't any others out there. If there were, chances are we would have been visited by now. Our galaxy, being about 100,000 light-years across and 12 billion years old, could have been traversed countless times by other intelligent civilizations. But it hasn't. Unless, of course, aliens have been here long before we have and didn't leave any clues, or do not want to make contact. (Taking the first 2 billion years off for good measure, and assuming intelligent civilizations can travel at least at one-tenth the speed of light, gives a total of 10,000 galaxy crossings in the last 10 billion years. Either we have been purposely ignored, or we are really inconspicuous.) Given the unknowns—how can we presume to understand an alien psyche if we don't even understand our own?—we should keep an open mind, repeating, as Carl Sagan suggested, that "absence of evidence is not evidence of absence." Maybe the aliens are just very shy.

If, indeed, we are a rare event, we must be ready to take on an enormous responsibility: we must preserve our legacy, learning how to survive in spite of ourselves. Humans are capable of the most wonderful creations and the most horrendous crimes. It is often very convenient to dream of archetypical aliens, wise and all-knowing, who will inspire and educate us before it's too late. Those aliens are not so different from the saints and prophets of many religions, who bring us hope and direction. But if we are alone, we must learn to save ourselves following our own guidance and acquired wisdom. It is here that a blending of science and religious ethics can be profoundly useful. We can start by extending the Old Testament maxim "Do unto others as you would have them do unto you" from society to all known and unknown living beings here and across the cosmos.

Then, we must learn from the way Nature operates. There is a single principle behind all existing order in Nature, an all-embracing urge to exist and to bind that manifests itself at all levels, from the racing world of subatomic particles to the edges of the observable Universe. It also manifests itself in our lives and our history. Humans cannot escape this alliance with the rest of the cosmos. Our tensions are part of this universal trend, our creations and destructions are part of the same rhythms that permeate the Universe. Through them, we search for transcendence, for a reality deeper and more permanent than our own. However, we have distanced ourselves from Nature and have become wasteful. Nature is never wasteful, it never uses more energy than it has to, it never chooses a more costly path to achieve the same end result. This is true of atoms, of bacteria, of elephants, and of galaxies. Our wastefulness is reflected in the way we treat our planet and ourselves. It is a cancer that grows and overwhelms what lives and what doesn't.

We must learn from Nature's simple elegance, from its esthetical and economical commitment to functionality and form. We must look beyond our immediate needs and greed, reintegrating ourselves into a physical reality that transcends political and social boundaries. Perhaps then we will start to respect our differences, to learn from those who believe differently than we do, who live and look differently than we do. And we don't have a minute to waste.

NATALIE ANGIER

Scientists Reach Out to Distant Worlds

FROM *THE NEW YORK TIMES*

Had some quorn lately? If the answer is no, then you probably are not planning on traveling to distant stars. The celebrated New York Times *reporter and best-selling author Natalie Angier explains why long-range space voyages are not likely to resemble life on board the USS* Enterprise.

Nobody knows why our early ancestors decided to get off their knuckles and stand upright. Maybe they just wanted a better view of the stars.

And when sky gazers finally realized that the heavenly lights were not the footprints of the gods, but rather millions of blazing stars like our Sun writ far, they began to wonder, How do we get there? How can we leave this world and travel, not merely the 238,000 miles to the Moon, or 35 million miles to Mars, but through the vast dark silk of interstellar space, across trillions and trillions of miles, to encounter other stars, other solar systems, even other civilizations? According to a group of scientists for whom the term "wildly optimistic dreamers" is virtually a job description, it will indeed be very difficult to travel to other stars, and nobody in either the public or private sector is about to try it anytime soon. But as the researchers see it, the challenge is not insurmountable, it requires no defiance of the laws of physics, so why not have fun and start thinking about it now?

At the annual meeting of the American Association for the Advancement of Science, held in February 2002 in Boston, scientists discussed how humans might pull off a real-life version of *Star Trek,* minus the space Lycra and perpetual syndication rights.

They talked about propulsion at a reasonable fraction of the speed of light, a velocity that is orders of magnitude greater than any spaceship can fly today, but that would be necessary if the light-years of space between the Sun and even the nearest star are ever to be crossed.

They talked about the possibility of multigenerational space travel, and what it might be like for people who board a spaceship knowing that they, their children, grandchildren and descendants through 6, 8 or 10 generations would live and die knowing nothing but life in an enclosed and entirely artificial environment, hurtling year upon year through the near-featureless expanse of interstellar space.

They talked about how big the founding crew would have to be to prevent long-term risks of inbreeding and so-called genetic drift. They talked about how the crew's chain of command would be structured, what language people would most likely speak, and what sort of marital and family policies might be put in place.

And they talked about food, all of which would have to be grown, cultivated and synthesized on board.

"One thing is almost certain," said Dr. Jean B. Hunter, an associate professor of biological and environmental engineering at Cornell. "You'll have to leave the steak, cheesecake and artichokes with hollandaise sauce behind."

Many of the subjects raised during the session were so fanciful that at times it felt like a discussion of how to clone a unicorn, and indeed half the presenters moonlight as science fiction writers.

Nevertheless, the researchers argued, human beings have shown themselves to be implacable itinerants, capable of colonizing the most hostile environments.

Dr. John H. Moore, a research professor of anthropology at the University of Florida, compared a theoretical crew of spacefaring pioneers to groups of Polynesians setting out tens of thousands of years ago in search of new islands to populate.

"Young people with food and tools would set out in large flotillas of canoes," he said. "Nobody knew if they would ever come back, the trade winds went in only one direction, and many of them perished in the ocean."

Yet over time, the Polynesians managed to colonize New Zealand, Easter Island and Hawaii.

Still, no human migration in history would compare in difficulty with reaching another star. The nearest stellar neighbor, the triple-starred Alpha Centauri, is about 4.4 light-years from the Sun, and a light-year is equal to almost 6 trillion miles. The next nearest star, Barnard's Star, is 6 light-years from home. To give a graphic sense of what these distances mean, Dr. Geoffrey A. Landis of the NASA John Glenn Research Center in Cleveland pointed out that the fastest objects humans have ever dispatched into space are the Voyager interplanetary probes, which travel at about 9.3 miles per second.

"If a caveman had launched one of those during the last ice age, 11,000 years ago," Dr. Landis said, "it would now be only a fifth of the way toward the nearest star."

Dr. Robert L. Forward, owner and chief scientist of Forward Unlimited, a consulting company that describes itself as "specializing in exotic physics and advanced space propulsion," argued that rockets and their fuel would be so heavy that they would prevent a starship from reaching the necessary velocity to go anywhere in a sane amount of time. He envisions a rocketless spacecraft that would be manufactured in space and equipped with an ultrathin, ultralarge sail, its span as big as Texas but using no more material than a small bridge. A beam of laser light or high-energy particles from a source on Earth, in space or perhaps on the Sun-drenched planet of Mercury would be aimed at the sail, propelling it and its attached module to as much as 30 percent the speed of light—or about 55,000 miles per second.

At that pace, said Dr. Forward, a crew would reach Alpha Centauri in under 50 years.

"You could get a bunch of 16-year-olds, train them and then send them out at the age of 20," he said. "They'd have a long, boring trip, reach Alpha Centauri when they're in their 60's or 70's, do some exploring, and send everything they learned back home."

Admittedly, the astronauts would not make it home themselves. "It's a lifetime job," Dr. Forward said. "But it could be done in a single generation."

For longer journeys, designed with multigenerational crews in mind, an onboard engine and fuel source would be required, perhaps something powered by nuclear bombs, or the combining of matter and antimatter in a reaction that converts both substances into pure energy.

However the ship is propulsed, the researchers agree that it must be comfortable for long-distance travel. That means creating artificial gravity by gently rotating the craft; a spin no greater than one or two revolutions per minute would suffice.

It might also mean calling upon architects with Disney-esque sensibilities.

"The inside of one of these long-duration space habitats might feel like the inside of a shopping mall," Dr. Landis said. "Malls are carefully designed to use space efficiently, yet to give you the feeling that they're more spacious than they are."

And malls, of course, are a great place to bring the family. In Dr. Moore's view, the good old-fashioned family is the key to success in space.

"Over the past several decades, space scientists and writers of science fiction have speculated at length about the optimum size and composition" of an interstellar crew, he said. They have imagined platoons of Chuck Yeager–type stalwarts grimly enduring all hardships, or teams of bionic and vaguely asexual crew members overseeing freezers of embryos that can be defrosted and gestated as needed.

"Some of the scenarios proposed so far are downright alarming from a social science perspective," Dr. Moore said, "since they require bizarre social structures and an intensity of social relationships which are quite beyond the experience of any known human communities."

In deciding how to organize a star mission, Dr. Moore looks to the most "familiar, ubiquitous, well-ordered and well-understood" of social forms, the human family. "Virtually every human society in history has been structured along kinship lines," he said, "from small-scale foraging societies to empires comprising millions of people."

Lines of authority and seniority in a family are reasonably clear, and when they're not, well, there's always the time-out chamber.

In Dr. Moore's rendition, all recruits for an interstellar odyssey would be guaranteed the opportunity, though not the requirement, to marry and have children. Mate choice would be part of the bargain as well, with the population cannily structured so that each cohort of individuals, on reaching sexual maturity, would have about 10 potential partners of a similar age to select from.

Dr. Moore and his colleagues have developed a computer simulation called Ethnopop, in which they asked how large the crew must be in order to maintain genetic variability over time while still allowing crew members a choice of sex partners. They determined that a founding crew could be as small as 80 to 100 people and stay viable for more than a thousand years, assuming that two rules were followed: women waited until they were in their mid-30's or so before having children, and they had only a couple each. Counterintuitive though it may seem, said Dr. Moore, delayed childbearing and small families are known to help maintain genetic variability in a closed population.

Genetic diversity may be essential, but Dr. Sarah G. Thomason, a professor of linguistics at the University of Michigan, argued that the same could not be said for language. "You want everyone to be able to talk to each other as soon as they're on board," she said.

As Dr. Thomason sees it, the likeliest lingua franca for a starship will be—*gracias a Dios*—English. After all, she said, English is the language of the international air traffic control system, the scientific community and the educated class generally. English is the official language of 51 of the 195 nations of the world, and it is the second language of many others.

Yet, while crew members will be expected to speak English, their accents are likely to be quite diverse, and the English that their children and grandchildren end up speaking will have a rhythm and texture of its own—Space English. And though Dr. Thomason believes that the basic structure of Space English is not likely to change much from that of the mother tongue, teenagers will, of course, invent words of their own and drop words of scant use. "I can imagine the loss of words like snow, rivers, winter, mosquitoes, if they're lucky," she said.

Another arena that will test the limits of human ingenuity is space cuisine. Without livestock on board or supply ships to restock the pantry, crew members will have to be entirely self-sufficient. Dr. Hunter of Cornell envisions crops grown in hydroponic gardens, in which plants are suspended in troughs like rain gutters, and water and fertilizer are trickled slowly over their roots. Among the possible food groups are wheat, rice, sweet potatoes, beans, soy, corn, herbs and spices.

In addition, space-minded agronomists are exploring the marvels of microbes. Plants take weeks to grow, but yeastlike microorganisms replicating in vats can be used to churn out significant quantities of carbohydrates, sugars, proteins and fats in a matter of hours. Of benefit to a community in which recycling is not just a personal virtue but a public necessity, microorganisms can live on the carboniferous waste products of plants and people.

"There's a protein product called quorn, which is made from filamentous mold," Dr. Hunter said. "Not to make a joke of it, but it does taste like chicken."

Some cliches, it seems, are truly universal.

MARGARET WERTHEIM

Here There Be Dragons

FROM *LA WEEKLY*

Astronomers, like magicians, perform with the aid of mirrors—specifically, the enormous mirrors that catch light from the edges of the universe. The distinguished science writer Margaret Wertheim tours a remarkable facility that manufactures these massive, yet exquisitely calibrated, windows to the stars.

Men of the Middle Ages sadly realized that the great dragons were long since gone from European soil. Only feeble remnants remained, paltry debased descendants of the grand saurians of the past: frilled snakes and lizards, and small, feathered, scaly-headed beasts not much bigger than a pheasant. The latter bore an uncanny resemblance to roosters, which had recently been imported from China and were still a bizarrity to European eyes. If the fearsome fire-breathing creatures of legend existed anywhere, it was in far-off lands at the edges of the known world. "Here there be dragons," the maps optimistically declared.

Historians Lorraine Daston and Katherine Park (authors of *Wonders and the Order of Nature*) alert us to a perverse tendency of wonders to congregate at the outer reaches of our cartographic knowledge. Throughout history, distant lands have beckoned with the promise of marvels: unicorns and elephants; giants, Cyclopes and races of dog-headed men; miraculous healing springs and trees whose gourds enclose, like fruits, miniature fleecy lambs. Distance loosens the mind, freeing the imagination from the restraints of common knowledge and opening the doors of perception to strange and unlikely counterintuitive phenomena.

Adventurous persons, from Marco Polo to Neil Armstrong, have always been willing to travel immense distances to experience wonders for themselves—expeditions have been mounted, novel conveyances constructed and fortunes expended. Today, of course, cartographic knowledge exceeds the bounds of our planet, and the domain of the marvelous has retreated, as it always will, to even farther fringes. These days, those in search of the preternatural look not across the Atlantic but beyond the horizons of geography itself, to what Kant called the "island universes" of distant galaxies. Ever since Galileo pointed his "optick tube" to the heavens and discovered mountains on the moon and "satellites" orbiting Jupiter, outer space has become our chief domain of marvels. Here there be dragons indeed: quasars spitting the energy of entire galaxies, cosmic strings thrumming with the original Primal Force, neutron stars so dense a teaspoon weighs as much as Everest, and black holes so powerful they could shred a spaceship into strings of spaghetti.

The ties that bind matter to space prevent us from voyaging in person to this fabulous frontier; absent a revolution in physics and a radical new form of propulsion, humanity seems destined to remain on our ancestral cosmic home. Miraculously, however, light is exempted from Einstein's laws, confirming perhaps the great physicist's belief that if "God is subtle, he is not malicious." Ephemeral and immaterial, light bears witness across the universe. Where adventurers past were propelled on ocean waves toward the lands of their dreams, so the phantasms of distant cosmic landscapes are borne to us across oceans of space on waves of light. Here, the wonders come to us, though again, Herculean effort is required for proper apprehension of the magical phenomena—which is why astronomers build telescopes.

Sometimes, bigger really is better. The speed and power of sailing ships depended on the size of their sails; so, the bigger the telescope mirror, the more light waves you can catch. Translating this into the metric of marvelousness—which, in opposition to gravity, *increases* with distance—the larger the mirror, the farther out into space you can see, and hence the more marvels you can behold. This tyranny of numbers was majesticaly brought home to me on a recent trip to the University of Arizona's Mirror Lab, where the world's largest telescope mirrors are made. There is nothing minimal about the place, which is in itself a haven of wonder.

Even before you enter the Mirror Lab, a touch of the surreal hovers about the enterprise, for it is bolted to the side of the university's sports stadium, the only structure on campus strong enough to support the huge machinery that casting requires. Inside it is more aerospace than bench top; the main workroom stretches three stories high and is half the size of a football field. Gigantic gantries crisscross the cavernous space, while massive cranes stand by with claws unclenched; they must be strong enough to heft 20 tons, yet gentle enough to handle crystal. The whole building is low-pressurized to protect the nascent mirrors from dust.

At the far end of the lab, some 50 yards away, an enormous mirror is being polished: 8.4 meters in diameter, it seems impossibly big yet indescribably delicate. With its deep concave surface smooth and glistening, and bathed in water to aid the buffing, it resembles nothing so much as a vast contact lens. Telescope mirrors are augmented eyes, and this one has 12 times the light-gathering surface of the Hubble Space Telescope. It is one of a pair intended for the Large Binocular Telescope currently being constructed on Mount Graham, in the Quinlan Range west of Tucson, which will soon be the world's most powerful optical instrument. Maximal vision demands that no bump on the mirror surface be larger than 100 nanometers (about 500 times narrower than a human hair): If the giant mirror being polished here were expanded to the size of North America, there would be no protrusion higher than 4 inches. The custom-designed robotic polisher crawling over the surface acts like a mechanical caterpillar nibbling away atoms at a time. Amazingly, it will be at its task 24 hours a day, seven days a week, for eight to 10 months.

Where conventional telescope mirrors are spherical, the Mirror Lab's are parabolic, the most efficient shape for focusing light. As early as the 17th century, Johannes Kepler perceived that one way to make a parabola was to rotate a bowl of liquid—under the force of gravity, spinning liquid naturally configures itself to this unique mathematical form. A few telescopes have employed this idea using rotating bowls of mercury, but that's a toxic way to view the stars. In 1980, Mirror Lab founder Roger Angel realized that Kepler's insight could be implemented with molten glass, if only you could keep the whole apparatus spinning while the glass cooled and set.

A physicist by training, Angel tells me that when insight struck, his understanding of the chemistry of glass was nil. Though English by birth, he repaired immediately to that great American laboratory, the backyard, where in a homemade kiln he fused together a couple of Pyrex custard cups—enriching both the future of astronomy and the noble tradition of domestic science.

———

ON THE DAY I visited the lab, a rare treat awaited. A brand-new mirror had just been taken out of the colossal new oven and was sitting on its pallet like a gigantic freshly baked cookie. Most of the glass is in a honeycomb structure, with just a thin layer on top that will be polished to form the actual mirror surface. Angel explains that the honeycombing gives the mirror strength while radically reducing the weight. Still, we're talking 21 tons of ultrapure borosilicate glass. The oven itself is a gargantuan steel contraption, bristling with bolts and snaking tubes 10 meters in diameter and 2.5 meters high. This apparatus rests on a base 3.5 meters high that spins the entire construction seven times a minute. In flight it resembles a giant whirling pressure cooker. Normally, thermal expansion would tear the mirror apart, and to guard against that catastrophe, the floor is lined with aluminum plates sitting on a bed of steel ball bearings that allow the mold to expand and contract as the glass heats and cools. When it's cooking, the oven reaches 2,120 degrees Fahrenheit, the heat of the Earth's mantle 50 miles down, and hot enough to melt rock into magma.

In the Age of Sony, when the little black box is king, there is something tremendously comforting about Large Scale Engineering, which reminds us, as we seem to need reminding, that there is (still) a physical world beyond the virtual flicker of our screens. Extending our vision monstrously, Angel's mirrors take us to the far edges of material awareness, to those distant domains where the cosmos dreams, and where matter and space disport themselves in contradiction to natural law. As always on the periphery, the real becomes marvelous and the marvelous becomes real.

Jennifer Kahn

Notes from a Parallel Universe

FROM *DISCOVER*

They call themselves maverick theorists. Scientists call them cranks. Members of the physics department at the University of California, Berkeley, keep every letter they receive from these often delusional outsiders in an archive they call the X-files. Sifting through them for a grain of truth, Jennifer Kahn gets a disorienting glimpse of a kind of Bizarro-world science.

Eleven years ago Eugene Sittampalam was sitting in a hotel room on the Libyan coast when he stumbled, as if by fate, on the unified field theory of physics. "I was on an engineering project at the time, with hardly any social life," he says. "I would retire to my room after dinner. I would switch on the radio, relax at my table, and start doodling." The problem that occupied him has stumped physicists from Albert Einstein to Stephen Hawking: how to join together the profound yet disparate insights of general relativity and quantum theory. But Sittampalam's doodling, apparently, drew connections that the rest had missed. "One thing led to another," he says, "and before the evening was over, I had the inverse square law of gravity derived—for the first time ever—from first principles!"

Sittampalam has no advanced degrees in physics. His theory is girded by mathematics no more complicated than high school algebra. Still, his claims are modest compared with those of other "maverick theorists," or cranks, as most scientists call them. At the American Astronomical Society meeting in

1999, a freelance astronomer argued strenuously that connecting certain pulsars across the night sky made an arrow that pointed directly to a vast alien communications network. A few years before, at Dartmouth, a dishwasher swamped the Internet newsgroups with his descriptions of the universe as a giant plutonium atom. The man, who identified himself as Archimedes Plutonium, wrote songs praising this atom universe and also provided stock tips. When he appeared on campus, it was in a parka covered with equations like a necromancer's robe.

Judging from the reams of odd theories sent daily to science journals, universities, and researchers, science cranks are more prolific than ever. This is true despite a discouraging silence on the part of the recipients. The author of one atmosphere-based theory of gravity estimates that he has mailed 5,000 copies of his work to physicists over the past 15 years but received just two replies. Presentation is part of the problem. "GENTLEMEN ARE YOU INTERESTED IN SEPARATING VALUABLE CHEMICAL COMPOUNDS FROM THE SUNSHINE RAY?" demands one impatient correspondent. Crank papers are so consistent in their tics that they're sometimes hung on physics department bulletin boards and given ratings—with points awarded for bold type, multiple exclamation marks, and comparison of self to Newton, Einstein, or God. But a few, like Sittampalam's, are more difficult to dismiss.

Sittampalam holds a bachelor of science degree from the University of Ceylon and has spent 20 years consulting for a number of prominent global engineering firms. His 85-page treatise is formatted with flawless professionalism, and he has no history of psychological disorders. Yet since his "breakthrough" in Libya, Sittampalam has all but sidetracked his career in pursuit of his theory. He has repeatedly sent his treatise to universities, paid to self-publish the work in paperback, and lost "a small fortune in salary" by his own estimation. Seven years ago he even offered a $25,000 reward to any physicist who could refute his theory and, as he puts it, "slap me out of this obsession." So far, no one has come up with a sufficient rebuttal.

Such single-minded absorption is part of the mythology of science. It's no wonder, then, that scientists are nearly as fascinated by cranks as cranks are by science. "It's unnerving," says Geoff Marcy, an astronomer at the University of California at Berkeley. "It shows how easy it is to slip from healthy, even necessary, conviction into certainty and delusion. Plus, you realize that you don't always know which camp you're in." There's the rub. Science owes a good part of its success to its capacity to contend with doubt—to engage it, respond to it, and transform itself in the encounter. Yet there's rarely a point at which a good idea becomes clearly, incontestably a bad idea. Neurologist Stanley Prusiner spent 15 years arguing that a misfolded protein called a prion caused the brain decay associated with scrapie and mad cow disease. Researchers snickered at him. Evidence slowly accumulated in his favor, and in 1997 he was awarded the Nobel Prize in medicine. "It's like a ball on top of a saddle," Marcy says. "You can't listen too closely to the establishment or you'll never be creative. But if you don't listen enough, you fall over the edge."

I FIRST CAME ACROSS Sittampalam's theory in the Berkeley physics department. There, for the past 20-odd years, the secretaries have diligently compiled what they call the X-files: the mother lode of crankiana. Kept in a three-foot-wide cabinet, the files contain hundreds of submissions, including one man's musical CD about thermodynamics and another's explanation of relativity and quantum mechanics spelled out on six postcards. Elsewhere on campus, researchers maintain what amount to branch libraries of the X-files. "I have an entire shelf of crank mail," MacArthur-winning physicist Rich Muller told me. "My favorite is a book written by a crank that includes all the letters she received from scientists."

Muller's office at Lawrence Berkeley Laboratory sits several hundred feet above the city, in a stolid cement building edged by eucalyptus trees. The lab's newly heightened security was in force, and I was allowed through the gate only after a lab employee turned up to vouch for my good intentions. When I arrived, Muller had everything laid out, fat folders of letters and textbooks stacked across half of a colleague's desk. "There was a poster of the universe," he mumbled, peering up at the room's highest shelf. "It was beautiful. I put it someplace special. Now I don't know where it is."

Superficially, Muller is a bit cranky himself. His hair is thin but mussed, and his office is a cave of overstuffed folders and yellowing articles tacked to a corkboard. He is the author, among other things, of the controversial Nemesis theory, which argues that a second sun caused the extinction of the dinosaurs, and a novel that explains some biblical miracles as clever but scientifically consistent sleight of hand. Muller corresponds with cranks and has thought enough about them to sort them into a fairly elaborate taxonomy. "The range . . . is quite broad," he says. At the top of his hierarchy are the merely misguided: retired engineers who have strayed from load-and-strain calculations into surmises about relativity. The bottom of the stack is hairier: the Mullerian estate of the super-crank. Some super-cranks are harmlessly delusional, others dangerously paranoid, but none are very good at listening—a trait that drives Muller bats. "You take the time to explain the mistake in their argument, and they just ignore the explanation," he says bitterly. "They don't realize how much time scientists spend coming up with ideas and rejecting them."

Cranks, of course, see it differently. In their view they are Davids fighting a Goliath. Sometimes their foes may be theorists who have gone too far ("Deception, horn-swoggling . . . Who are you fooling?" demands an opponent of string theory). Other times they are scientists—overeducated, institutionalized, hidebound—who don't dare go far enough.

This confusion over fundamental purpose is understandable, given that modern physics manages to seem at once simple and profoundly puzzling. Astronomers have only recently determined that a mysterious "dark energy" is forcing the universe apart, overwhelming the equally mysterious "dark matter" that seemed to be holding it together. Even gravity, faithful shepherd of falling rocks and fly balls, has recently gone to pieces: At small distances, it may not be constant at all. "Some of the ideas are incredibly counterintuitive," says Nima Arkani-Hamed, a Harvard physicist who specializes in theoretical particle physics. "And they're just getting more bizarre."

Arkani-Hamed himself believes that space contains seven extra dimensions we can't see because they're rolled up like very small window shades. His mannerisms, too, might seem suspect in someone with less impressive credentials. He talks faster than I can take notes, a kind of super-revved speech that still seems to fall frustratingly short of the speed of thought. "Certain traits of personality and character are . . . close," he admits. "The obsessive tendencies, the compulsion, the restlessness. It's not the same, but there's a resemblance." Then he adds, dryly: "A lot of scientists have traits that would be bizarre if not channeled into science. I know that's part of why cranks interest me."

AFTER SEVERAL DAYS of reading the X-files, I felt as if I were attending school in a parallel universe. "It is imperative that we begin burning water as fuel!" one author urged. Others were more puzzling. A note written on a ripped sheet of notebook paper said only, "I contend the holes on the right side of these pants are not explainable by contemporary science." A few submissions aped the style of scholarly papers, including credentials: An outline for "Symmetrical Energy Structures in a Megadimensional Cosmology," for instance, came from the director of the Alpha Omega Research Foundation in Palm Beach, Florida. But most favored a more urgent style. Arguments crescendoed to uppercase type. Words, boxed and colored, squeezed together on the page like castaways on a homemade raft.

At times the grandiloquence was so ingenuous it was hard to hold much of a grudge. "Readers, stretch your imagination to the very limits!" the inventor of Wavetron theory implored. "Together we will batter back the barbarous hordes!" The boldface words in another paper, taken together, read nearly like verse: "The eye is low / Negative ground / Electricity compressed, dead calm, displacing space / No one knows the cause / displacing . . . / repelling . . . / Well I do." But not every crank is so poetic nor so benign. Arkani-Hamed described one author whose e-mails had become increasingly virulent. Another physicist refused to be quoted by name in this article, replying tersely: "There is no guarantee that all cranks are harmless." Still another described his feelings about cranks as "Neutral. With a touch of fear."

One case in particular has echoed down the years with the force of a small-town murder. In 1952 a man named Bayard Peakes turned up at the office of the American Physical Society at Columbia University with a gun. Peakes was frustrated at the society's rejection of his pamphlet, "So You Love Physics." Unable to find any physicists at the society's office, he shot and killed a secretary instead. (Just months before, ironically, the society had changed its policy to open its annual meetings to public speakers and accept *all* scientific abstracts—including another by Peakes that aimed to prove that the electron doesn't exist.)

The Peakes case was unique in degree but not in kind. Scientists have been heckled, cursed, and harassed at work (one crank faxed love letters to a department chair and forged the signature of another scientist at the bottom). A few have even had cranks turn up at their homes.

It was hard not to have these cases in mind when I began contacting writers from the X-files, using the information that came with some of the papers. For the most part the authors were elusive. Phones had been disconnected, e-mail addresses bounced. The few who did answer were single-minded. One retired commercial diver answered all my questions with an uninterruptible monologue on gravity (it pushes rather than pulls, he said). An elderly man in southern California called back half a dozen times, each time hinting at his latest discovery.

"With psychosis, there's a kind of pressure to push it out," John MacGregor, an expert in the "outsider art" produced by mental patients, told me. "Sometimes the manic-depressives don't even use periods. They don't want to stop writing!" The trouble starts when such zeal is spiked with paranoia. "Schizophrenics have a tremendous desire to prove that they're sane," MacGregor said. "It could be that they've adopted science in order to prove just how rational and intelligent they are." He paused. "If a paranoid schizophrenic decides that certain rays are emanating from the physics department, it could be dangerous. These are the people who might come in and shoot it up."

Compared with the people MacGregor described—even compared with some of the physicists I interviewed—Sittampalam was charming. On the phone from his home in Sri Lanka, he proved candid but not overbearing, with crisp, British-inflected English pleasantly free of run-on tendencies. He answered questions about his family (he has five brothers and has never married) and chatted easily about his current job at ElectroFlow, a Missouri-based start-up that helps companies optimize their power consumption. He maintained that his physics theories were quite accessible; indeed, he hoped to see them introduced at the high school level.

I liked Sittampalam enough to inveigle a physicist friend to read Sittampalam's paper, with the promise that he remain anonymous. I was secretly hoping the paper would have some merit, or if not, that it would contain a clear error: one that, recognized, would set Sittampalam free from his compulsion. But when my friend got back to me, the news was bad. "As I read this, I kept thinking: 'How hard can it be to prove that this paper is incontrovertibly wrong?' " he said. "But it is hard. Not because his ideas are right. They're not. But because he's created a self-consistent system of arguments."

Self-consistency is not in itself a valuable trait—the theory that aliens created Earth and continue to control its evolution is a self-consistent system— but it can make things hard to refute. "I'd love to find just one equation in here and say, 'We have observations proving that's not correct,' " the physicist said. "But there's no mathematical progression. He starts with some very basic equations from classical mechanics. He mixes, stirs, spends some time hypothesizing in a very general way about physics, and out pops another familiar equation: $E=mc^2$. But really, he's just waved his hands. He could never have gotten to that next equation if he didn't already know what it was—and he knew what it was only because other people had figured it out for him using the traditional framework of physics."

Reading Sittampalam's paper feels a bit like being in a hedge maze: Just when you think you're heading toward some grand, central idea—an explanation of the cosmological redshift, for instance—the discussion loops away for another, more distant destination. There is the matter of Earth, for example. Sittampalam claims that his theory is the only way to explain why Earth hasn't lost enough energy over the years to spiral into the sun. But a physicist who saw the paper wrote in to note that that's exactly what *will* happen—just billions of years from now. Sittampalam acknowledged that mistake but attributed it to a typo. He had mistakenly left the words "under perturbation" out of his hypothesis, he said. Revised, his theory now explained why Earth, subject to the gravitational pull of the rest of the planets, has never wandered out of its orbit.

"First, he's talking about gravitational radiation, which is a real but minute effect; now he's talking about the solar system being sensitive to small changes," the physicist said. "It's true that if you moved the Earth a little bit today, its position and velocity in a month would become quite different. But that doesn't mean the shape of the current orbit is going to fall apart. We have simulations showing just the opposite, actually: that the solar system is stable over an incredibly long timescale. But that's what I mean. Every error you find, he's just going to change the subject. It's never-ending."

THE TRUTH, dispiriting as it may seem, is that cranks are pretty much never right. "We'd love it if one of these guys were right," Arkani-Hamed says. "A revolutionary idea that works—great!" But real science tends to advance by increments rather than by revolutions. The life of working scientists is long on tedium and short on glory. They write grants, sit on committees, do paperwork. There is pressure to play it safe and be competitive. Cranks, by contrast, are free agents. With no career to lose and no scientific framework to restrict them, they can publish at their own pace and dare to shoot for the moon.

All of which may explain why most cranks aren't scientists and presumably wouldn't want to be. It may also explain why some scientists, when they talk about cranks, evince something close to envy. "There's curiosity, excitement, a kind of purity of purpose," Geoff Marcy says. Unlike conspiracy theorists, science cranks inhabit a happy universe: one that's accessible to those who plumb it ("Dear universal adventurer!" one postcard about quantum gravity begins). To read their ideas is a vicarious thrill, Arkani-Hamed admits, "but eventually you go back to what you were doing. In the end, the thing that makes science so amazing is that it works."

As for Sittampalam, he suspects that the poor reception for his work is largely a political matter. "I can easily answer all the critical points he raises," he replied, when I forwarded the physicist's critique. "But will he be convinced?" In the preface to his thesis, Sittampalam quotes Sir Martin Rees, a renowned astrophysicist and Astronomer Royal at Cambridge University. "Generally, researchers don't shoot directly for a grand goal," Rees writes. "Unless they are geniuses (or cranks) they focus on problems that seem timely or tractable."

When I asked Sittampalam which he is, genius or crank, he was surprisingly equivocal. "Perhaps I'm a crank, but that's left for history," he said. "I have no regrets. When your work is for the future, by necessity you are not understood in your own days."

In the meantime, he can take comfort from the case of the Indian mathematician Srinivasa Ramanujan. In 1913 Ramanujan was a clerk at Madras Port Trust—"a short uncouth figure," in the words of one contemporary; "stout, unshaven, not over clean, with one conspicuous feature: shining eyes." Although largely self-taught in mathematics, Ramanujan had the audacity to mail 120 of his theorems to the British mathematician Godfrey Hardy at Cambridge University. Hardy dismissed the pages as gibberish at first, only to find, upon careful consideration, that some of the theorems were truly revelatory. Five years later Ramanujan was elected to the Royal Society of London.

MICHELLE NIJHUIS

Shadow Creatures

FROM *HIGH COUNTRY NEWS*

First Yuppies. Now Yuckies—"Young Urban Crows." With suburban sprawl displacing animal habitats, metropolitan areas aren't so much pushing nature out as creating new niches for wildlife—and vexing problems for municipal officials and environmentalists alike. The solutions aren't always easy, as Michelle Nijhuis reports from the Seattle suburbs, where the crows are practically the fastest-growing segment of the population.

It doesn't seem too difficult to trap a crow. Especially if you're armed with a remote-controlled, rifle-powered, 25-foot-square net and a heap of stale white bread. Especially if you've seen the crow in question almost every day for the past six years. Especially if it lives just a couple of wingflaps from your own suburban backyard.

It's harder than you might think.

"Bastard!" explodes John Marzluff, an otherwise even-tempered wildlife biologist from the University of Washington. He tosses the remote control for the net gun on the dashboard of his truck and tries to take a deep breath.

For the second time on this gray, low morning, he's pushed the button on the remote, and for the second time exactly nothing has happened. No net has shot out of the ferny underbrush, no panicked crow is struggling for freedom, no one is running forward to fit the bird with an identifying leg band. Instead, less than 100 yards down the conifer-edged road, a female crow is strutting well within range of the stalled net, stuffing her beak with bread.

Marzluff has spent his career studying crows and ravens in Arizona, Maine, Idaho, Montana, Hawaii and Guam. He and his students have banded about 500 crows in the Seattle area, but the job doesn't get much easier with practice.

"The more you try to trap crows," he sighs, "the shorter your lifespan."

Crows and their cousins in the corvid family, ravens, jays and magpies, have spent hundreds of thousands of years taking advantage of our inventions. Today, they forage in dumps and on suburban lawns; they follow hunters to prey and backpackers to campsites; they nest on Alaskan oil rigs and in the ornate stonework of city libraries. They've been known to perform pitch-perfect imitations of explosions, revving motorcycles and flushing urinals.

They're fiercely, exasperatingly smart.

It's all too easy for crows to survive in the Seattle suburbs, where they have free access to truckloads of tasty human castoffs. While many species are forced to flee the expanding rings of development, crows and a few other hardy creatures are rushing in like bargain-hunters on their way to a flea market. Like it or not, our backyards are hosting an evolutionary showdown, and the odds favor the coyotes and the crows: The coyote is the only midsize carnivore that is actually expanding its range in North America; the American crow, once rare in the Pacific Northwest, is now one of the dominant birds in the Seattle area.

The showdown is pressing many Seattle residents—and the rest of us—up against an awkward truth. Though we might like our cities neatly separated from the natural world, nature is having none of that. Wild animals are reacting and adapting to us as fast as they can, not just to our logging and mining and ranching and fishing, but also to our fast-food restaurants, golf courses and campgrounds.

Marzluff and a few of his colleagues are proving as adaptable as the animals they study. In recent years, they've moved their research out of the wilderness and into the suburbs. By shadowing the animals that shadow us, they're discovering how we might protect other, less adaptable creatures from being elbowed out by the flood of newcomers.

"Crows are a perfect mirror for us," says Marzluff. "They're a good species for people to look at, not because crows are doing something wrong, but because we're doing a lot wrong—and they're taking advantage of it, every step of the way."

Ever since the late 1800s, when Seattle was little more than a staging ground for the Klondike gold rush, the city has had an irony-laden relationship with wildlife. Even then, city boosters were promoting Seattle as nature's next-door neighbor, a place that provided a quick escape from the distractions of urban life. Seattleites were also doing their damnedest to control the natural processes around them, and they dug waterways and filled tidelands as busily as any beavers.

Despite boosters' best efforts, wildlife refused to cooperate. Muskrats undermined a dam in central Seattle in the early 1900s, causing major damage to the Fremont Bridge, and so many frogs filled a canal near the Duwamish River that residents feared for local water quality. In the 1930s, city park officials encouraged the feeding of birds, hoping to please nature-loving visitors, but the mobs of geese and other waterfowl polluted Green Lake with droppings, uprooted flowers and shrubs, and created an uproar among local residents.

By the 1990s, the city had transformed itself. It was the hippest spot on the West Coast, with a Microsoft-powered economy, a caffeinated sensibility and an influential downtown music scene. More than half a million people moved to the area during the decade, many of them young, college-educated and eager to be nature's neighbors.

Instead of the peaceful, outdoorsy life they envisioned, the newcomers encountered some very urban problems, including a desperate housing crunch and some of the worst traffic tangles in the country. They also encountered crows—lots of them.

Suburban housing developments and landfills "are like a banquet" set especially for crows, says Marzluff. "We're creating hundreds of acres of crow habitat every single day," he says. "We're creating habitat faster than the crows can fill it."

Like humans, crows tend to breed in the food-rich suburbs. Juveniles without established territories spend more time in the poorer habitat downtown, moving back into the 'burbs when they find mates. (Marzluff and his students, who track the movements of their banded and radioed birds, call these adolescent wanderers the Young Urban Crows, or "yuckies.")

This survival strategy has been a wild success: The area's crow population has grown by as much as tenfold in the past two decades, and it grew by more than 30 percent just last year. It's one of the fastest-growing crow populations in the world, and the birds are getting hard to ignore.

Crows peck at mossy cedar shingles, drink from gutters and find their way into downtown office buildings. Karen Rillo and Mike Mead, the owners of a nuisance-wildlife franchise called Critter Control, are on the receiving end of many of the resulting complaints. They've shooed a crow out of a Barnes and

Noble in University Village, used reflective balloons to scare crows off rooftops, and advised sleepless homeowners to spook the birds by hanging a dead crow in a tree. But the noisy flocks often prove persistent.

"I'm taking it personally when five pillows over my head won't do the trick," Seattle resident Susan Brett told the *Seattle Times*. After a night of tossing and turning, she said, she spent the morning looking at newspaper ads for air guns.

Matthew Klingle, a history professor at Bowdoin College in Maine who wrote his Ph.D. dissertation on the environmental history of Seattle, says such conflicts haunt almost every city in the United States—and are particularly persistent in the West.

"People think about the West as nature's province, and they move to Seattle, Portland, Boise or Salt Lake City to be close to nature," he says. "But people also want clear boundaries. They want a divide between nature and culture."

Crows aren't the only animals causing headaches for their human hosts, and Seattle isn't the only city that's unintentionally making more and more room for crafty wildlife.

In Phoenix, hungry javelinas—knee-high wild pigs—can't resist the exotic landscaping in suburban yards. "I tell people that they're just putting an ice cream parlor on their corner," says Arizona Game and Fish wildlife biologist Joe Yarchin. His office, which handles more than 1,000 nuisance-wildlife complaint calls every year, deals with Gila woodpeckers that hammer at air-conditioning units, peregrine falcons that smear pigeon guts on downtown law-office windows, and most everything in between.

His typical call, though, has something to do with coyotes.

Like crows, coyotes have long been associated with humans. They're our companions and our guides, our jesters and our harassers in legends and myths. And also like crows, coyotes are having a high time in the suburbs. In recent decades, their populations have rebounded from the all-out extermination efforts in the first half of the 20th century, and they've started Dumpster-diving around urban parks and suburban backyards.

Coyotes tend to keep a low profile. Though the Game and Fish office in Phoenix gets a lot of complaints about coyotes every year, not many of the animals are really causing any trouble. But during the painfully dry summer of 2002, a family of seven coyotes kept turning up in a tony Phoenix neighborhood; a group of skinny juvenile coyotes was seen hunting ducks in a suburban park; and more than a few cats and dogs came home with telltale battle scars.

Coyotes have also made themselves at home in Tucson, San Diego, and Denver. They're regularly spotted in Oakland, California, and South San Francisco, and U.S. Geological Survey biologist Erin Boydston began to track several recently arrived packs in Golden Gate Park. In Portland, Oregon, surprised public-transit employees found a coyote inside a city light-rail train, calmly curled up on a seat. The incident even inspired a song, "Light-Rail Coyote," an ode to Portland by the ultra-popular band Sleater-Kinney.

Of course, coyotes have moved into Seattle, too. Biologist Timothy Quinn, whose dissertation research on urban and suburban coyote behavior sent him striding down Seattle sidewalks with a radio receiver, heard reports of coyotes in the Woodland Park Zoo (where they were trying to eat some frightened peacocks) and in heavily visited Discovery Park on the edge of Puget Sound. Several years ago, a young coyote wandered into a downtown office building, where wildlife officials cornered it in an elevator. "That was one scared coyote," Quinn remembers.

Quinn, now the chief scientist of the habitat program for the Washington Department of Fish and Wildlife, says the suburbs are as much of a banquet for coyotes as they are for crows. When Quinn was collecting coyote scat for diet analysis, he walked the same routes every two weeks. "I always saw all these little cat collars . . . at first, it didn't make sense to me," he says.

His analysis eventually showed that coyotes' single most important mammalian prey was the suburban housecat. Then, he says, "all those little collars started to make sense."

Biologists aren't sure how these urban and suburban coyotes affect their ecosystems, or how quickly changing, human-dominated ecosystems affect coyote behavior. Quinn says coyotes in Seattle might be a boon to songbird populations, since they pick off so many warbler-stalking kitties, but he can only guess.

JOHN MARZLUFF'S SECOND crow-trapping stop of the morning is in a new subdivision, one packed with trimmed lawns, hopeful landscaping and cedar-shingled three- and four-bedroom homes. As we pull over to the curb and hop out, a sprinkler near our feet starts up with a sudden *pfft*.

Marzluff sets up his net gun, and we quietly settle in for another wait. Almost immediately, a flock of juvenile crows starts cawing on the next corner, and soon a small group of them begins circling the hill of white bread. Marzluff leans forward, remote control in hand, and—yes!—the blank rifle cartridges explode, the net soars out, and one young crow is stopped in its tracks.

A woman in a tailored black suit and heels pokes her head out of the nearest house, taking in the truck, the biologist and the unlucky crow. "What was that?" she demands. Marzluff explains and apologizes, and the woman shrugs, her curiosity satisfied for the moment.

Most biologists don't have to consider the effects of nervous neighbors, speeding cars or ill-timed landscaping work. They've long preferred to work in big nature, in wilderness areas and other places where nature's gears turn in relative peace. For decades, many have viewed cities as ecologically dead, places where natural processes stalled out long ago.

Marzluff likes studying the suburbs, not just because he's fascinated by the ingenuity of crows ("You get hooked on 'em," he says) but also because he's trying to figure out how other, less-adaptable species get by in the sea of subdivisions.

The total transformation of this landscape, along with the crows' habit of aggressive nest predation, should be a death sentence for any forest-loving animal. But in the struggle between the garbage-eaters and the habitat purists, some of the purists are turning out to be surprisingly tenacious.

Just a few hundred yards down the wide, curving road, a slender greenbelt snakes around the edge of the development. This tiny area, barely 45 yards wide and just over a mile long, is an unexpectedly effective wildlife refuge. Though the number of birds isn't nearly what it would be in an undisturbed stretch of forest, every feathered forest-specialist in the region has appeared here at one time or another. From the well-established trail, Marzluff points out a winter wren nest, a delicate, grapefruit-sized ball of moss.

This smidgen of forest may not be attractive habitat for long. Invasive plants may creep in, or curious cats and kids may disturb nesting patterns. But for now, the greenbelt is like an island with regular ferry service to the mainland. With a 150-acre University of Washington forest preserve just down the road, wrens and other birds can usually find the food, mates and habitat they need by traveling between the two areas. Marzluff and his colleagues at the University of Washington's Urban Ecology program have found that such well-managed small areas, interspersed with larger preserves, could go a long way toward maintaining stable populations of forest birds and other animals.

These hopeful results are probably a happy accident, since parks and green spaces are most often designed for us, not for wildlife. Parks are intended, overtly or not, to educate us, enlighten us, or entertain us; animals, if they appear, are usually just a pleasant diversion for passersby. Marzluff hopes his work will convince some planners to take a bird's-eye view.

"We don't want to just set aside habitat, we want to set aside functional habitat," he says. "We want to make sure we have a good mixture, that it's not all low-density sprawl."

He and a few other researchers argue that cities and other human-dominated places are far from dead environments. They say they're complex ecosystems, constantly in flux and well worth the attention of a new generation of ecologists. They hope to flush more of their colleagues out of the woods to investigate, and they're getting some high-profile support.

The federally funded National Science Foundation, which underwrites the work of the University of Washington's Urban Ecology program, also oversees a network of about 20 long-term ecological research stations. In 1997, the foundation chose Baltimore and Phoenix for its first urban research stations. The Phoenix station currently supports more than 50 projects, and many involve not only biologists but also economists, sociologists and urban planners.

Through her work at the Phoenix station, Arizona State University biologist Ann Kinzig has found that desert birds can also take advantage of habitat fragments in the city. Small neighborhood parks—"even places with playgrounds and baseball fields"—support rich populations of native birds, ones that seem to coexist with human-associated species such as rock doves and starlings. When she applied economic and demographic data to her findings, she discovered that bird diversity is significantly higher in wealthier neighborhoods, a tantalizing pattern she plans to investigate further.

The field of urban ecology still has a long way to go. "This research is where timber research was 20 years ago," says Andrew Hansen, an ecology professor at Montana State University who studies the impacts of rural subdivisions. In the early 1980s, he says, biologists knew very little about the effects of clear-cutting, but the lengthening roster of endangered forest species inspired concentrated research.

"We learned a lot about how the ecosystem worked, and we were able to figure out how to log more gently," he says. "We're just now realizing that rural and urban development is a serious issue in many areas. We're just beginning to come up with ways to live more lightly on the land."

In 20 years, this research will be even more critical. A recent study in the journal *Bioscience* reported that sprawl is already the top cause of species endangerment in the continental United States. In July 2002, the American Farmland Trust estimated that 25 million acres of Western ranchland will be threatened by low-density development within the next two decades.

"Pretty soon," says Tim Quinn, "we're all going to be urban biologists."

Cities might offer fascinating ecological puzzles to a new breed of scientists, but is "living more lightly on the land"—especially in Seattle or Phoenix—really worth the trouble of finding out how to do it? After all, most wildlife habitat in our cities has been more or less permanently paved over, and what little is left seems to be dominated by crows, coyotes and hungry wild pigs. It's hard not to see our backyards as sacrifice zones. Even John Marzluff, who can see hope in 24 acres of scraggly conifers, isn't always optimistic.

"Studying urban ecology makes you a fan of the timber industry," he says flatly. "The amount of disturbance we create where we live makes all the other environmental issues we have pale in comparison."

But in the modern metropolis, even small conservation victories can be meaningful. In Seattle, as in most urban areas, humans have built on top of high-value wildlife habitat (low-elevation valleys and coastal areas) and preserved more scenic but less diverse areas (mountaintops and ridges).

"People think the wildlife is out there, in the national parks," says John Kostyack, head of the smart-growth and wildlife program at the National Wildlife Federation. "Contrary to popular belief, we've found that [the suburbs] are quite rich wildlife areas." His group is delving into the environmental records of U.S. metropolitan areas, and released a report and a set of recommendations at the end of 2002.

Other national environmental groups, including the Natural Resources Defense Council and the Sierra Club, have established sprawl-control programs that include habitat-protection efforts. Many land trusts, most notably The Nature Conservancy, focus on protecting privately owned wildlife habitat instead of generic open space. Land trusts of all sizes are using conservation easements to protect wildlife-friendly lands on the urban fringe.

Thanks to the Endangered Species Act, some city officials are also getting into the business of habitat protection. Seattle has limited logging and altered flows in the Cedar River watershed to protect the endangered Puget Sound chinook salmon. Tucson, like San Diego before it, is embroiled in a massive habitat-conservation planning process triggered by a suite of troubled species.

In Seattle, the city-funded Urban Creeks Initiative has brought some of the poorest neighborhoods into closer contact with their nearby rivers, and transformed what were once seen as dumps and drainage ditches. Over the past decade, community groups on the southern end of Seattle have restored a peat bog and chopped out invasive plants in Longfellow Creek, while city agencies have piled up woody debris to slow down flows and make the waters more hospitable to salmon. In 2002, about 300 salmon came up the creek, among them a pair of Puget Sound chinooks.

Part of Longfellow Creek still runs under a Kmart parking lot, but many stretches are more accessible and more familiar to the whole community, says creek watershed specialist Sheryl Shapiro. "People are just astonished," she says.

"They'll say 'Hey, I've lived here all this time, and I never even knew this was here.' "

In her book *Flight Maps: Adventures with Nature in Modern America,* historian Jennifer Price writes that "we have used a very modern American idea of Nature Out There to ignore our ravenous uses of natural resources." We have tried very hard to remain, as Woody Allen put it, "two with nature."

The Longfellow Creek Project and other community efforts suggest we may be able to treat our neurotic relationship with the natural world. Perhaps we can turn capital-N Nature into something we successfully coexist with every day.

The crows and coyotes might lend us a hand here. They might be able to, in an odd, roundabout way, help us solve the problem they represent. By busting through our comfortable ideas about where the city ends and Nature begins, our annoying, overbearing wild shadows might finally convince us that how and where we choose to live has a lot to do with the future of the natural world.

They make our options crystal clear: We can continue making endless habitat for crows and their adaptable colleagues, or we can try to make enough room for everybody. It's up to us.

GUNJAN SINHA

You Dirty Vole

FROM *POPULAR SCIENCE*

What does a rodent have to tell us about love? Researchers studying brain chemistry and animal behavior have found that the humble prairie vole exhibits traits familiar to humans—mating and cohabitating to raise children, while occasionally straying for fleeting encounters with other partners. Gunjan Sinha has the latest findings from the science of love.

George is a typical Midwestern American male in the prime of his life, with an attractive spouse named Martha. George is a devoted husband, Martha an attentive wife. The couple has four young children, a typical home in a lovely valley full of corn and bean fields, and their future looks bright. But George is occasionally unfaithful. So, occasionally, is Martha. No big deal: That's just the way life is in this part of America.

This is a true story, though the names have been changed, and so, for that matter, has the species. George and Martha are prairie voles. They don't marry, of course, or think about being faithful. And a bright future for a vole is typically no more than 60 days of mating and pup-rearing that ends in a fatal encounter with a snake or some other prairie predator.

But if you want to understand more about the conflict in human relationships between faithfulness and philandering, have a peek inside the brain of this wee rodent. Researchers have been studying voles for more than 25 years, and they've learned that the mating behavior of these gregarious creatures uncannily resembles our own—including a familiar pattern of monogamous attachment: Male and female share a home and child care, the occasional dalliance notwithstanding. More important, researchers have discovered what drives the animals' monogamy: brain chemistry. And when it comes to the chemical soup that governs behavior associated with what we call love, prairie vole brains are a lot like ours.

Scientists are careful to refer to what voles engage in as "social monogamy," meaning that although voles prefer to nest and mate with a particular partner, when another vole comes courting, some will stray. And as many as 50 percent of male voles never find a permanent partner. Of course, there is no moral or religious significance to the vole's behavior—monogamous or not. Voles will be voles, because that's their nature.

Still, the parallels to humans are intriguing. "We're not an animal that finds it in our best interest to screw around," says Pepper Schwartz, a sociologist at the University of Washington, yet studies have shown that at least one-third of married people cheat. In many cases, married couples struggle with the simple fact that love and lust aren't always in sync, often tearing us in opposite directions. Vole physiology and behavior reinforce the idea that love and lust are biochemically separate systems, and that the emotional tug of war many of us feel between the two emotions is perfectly natural—a two-headed biological drive that's been hardwired into our brains through millions of years of evolution.

No one knew that voles were monogamous until Lowell Getz, a now-retired professor of ecology, ethology, and evolution at the University of Illinois, began studying them in 1972. At the time, Getz wanted to figure out why the vole population would boom during certain years and then slowly go bust. He set traps in the grassy plains of Illinois and checked them a few times a day, tagging the voles he caught. What surprised him was how often he'd find the same male and female sitting in a trap together.

Voles build soft nests about 8 inches below ground. A female comes of age when she is about 30 days old: Her need to mate is then switched on as soon as she encounters an unpartnered male and sniffs his urine. About 24 hours later, she's ready to breed—with the male she just met or another unattached one if he's gone. Then, hooked, the pair will stick together through thick and thin, mating and raising young.

Getz found vole mating behavior so curious that he wanted to bring the animals into the lab to study them more carefully. But he was a field biologist, not a lab scientist, so he called Sue Carter, a colleague and neuroendocrinologist. Carter had been studying how sex hormones influence behavior, and investigating monogamy in voles dovetailed nicely with her own research. The animals were small: They made the perfect lab rats.

The scientific literature was already rich with studies on a hormone called oxytocin that is made in mammalian brains and that in some species promotes bonding between males and females and between mothers and offspring. Might oxytocin, swirling around in tiny vole brains, be the catalyst for turning them into the lifelong partners that they are?

Sure enough, when Carter injected female voles with oxytocin, they were less finicky in choosing mates and practically glued themselves to their partners once they had paired. The oxytocin-dosed animals tended to lick and cuddle more than untreated animals, and they avoided strangers. What's more, when Carter injected females with oxytocin-blocking chemicals, the animals deserted their partners.

In people, not only is the hormone secreted by lactating women but studies have shown that oxytocin levels also increase during sexual arousal—and skyrocket during orgasm. In fact, the higher the level of oxytocin circulating in the blood during intercourse, the more intense the orgasm.

But there's more to vole mating than love; there's war too. Male voles are territorial. Once they bond with a female, they spend lots of time guarding her from other suitors, often sitting near the entrance of their burrow and aggressively baring their beaver-like teeth. Carter reasoned that other biochemicals must kick in after mating, chemicals that turn a once laid-back male into a territorial terror. Oxytocin, it turns out, is only part of the story. A related chemical, vasopressin, also occurs in both sexes. Males, however, have much more of it.

When Carter dosed male voles with a vasopressin-blocking chemical after mating, their feistiness disappeared. An extra jolt of vasopressin, on the other hand, boosted their territorial behavior and made them more protective of their mates.

Vasopressin is also present in humans. While scientists don't yet know the hormone's exact function in men, they speculate that it works similarly: It is secreted during sexual arousal and promotes bonding. It may even transform some men into jealous boyfriends and husbands. "The biochemistry [of attachment] is probably going to be similar in humans and in [monogamous] animals because it's quite a basic function," says Carter. Because oxytocin and vasopressin are secreted during sexual arousal and orgasm, she says, they are probably the key biochemical players that bond lovers to one another.

But monogamous animals aren't the only ones that have vasopressin and oxytocin in their brains. Philandering animals do too. So what separates faithful creatures from unfaithful ones? Conveniently for scientists, the generally monogamous prairie vole has a wandering counterpart: the montane vole. When Thomas Insel, a neuroscientist at Emory University, studied the two species' vasopressin receptors (appendages on a cell that catch specific biochemicals) he found them in different places. Prairie voles have receptors for the hormone in their brains' pleasure centers; montane voles have the receptors in other brain areas. In other words, male prairie voles stick with the same partner after mating because it feels good. For montane voles, mating is a listless but necessary affair, rather like scratching an itch.

OF COURSE, human love is much more complicated. The biochemistry of attachment isn't yet fully understood, and there's clearly much more to it than oxytocin and vasopressin. Humans experience different kinds of love. There's "compassionate love," associated with feelings of calm, security, social comfort, and emotional union. This kind of love, say scientists, is probably similar to what voles feel toward their partners and involves oxytocin and vasopressin. Romantic love—that crazy obsessive euphoria that people feel when they are "in love"—is very different, as human studies are showing.

Scientists at University College London led by Andreas Bartels recently peered inside the heads of love-obsessed college students. They took 17 young people who claimed to be in love, stuck each of them in an MRI machine, and showed them pictures of their lovers. Blood flow increased to very specific areas of the brain's pleasure center—including some of the same areas that are stimulated when people engage in addictive behaviors. Some of these same areas are also active during sexual arousal, though romantic love and sexual arousal are clearly different: Sex has more to do with hormones like testosterone, which, when given to both men and women, increases sex drive and sexual fantasies. Testosterone, however, doesn't necessarily make people fall in love with, or become attached to, the object of their attraction.

Researchers weren't particularly surprised by the parts of the lovers' brains that were active. What astonished them was that two other brain areas were suppressed—the amygdala and the right prefrontal cortex. The amygdala is associated with negative emotions like fear and anger. The right prefrontal cortex appears to be overly active in people suffering from depression. The positive emotion of love, it seems, suppresses negative emotions. Might that be the scientific basis for why people who are madly in love fail to see the negative traits of their beloved? "Maybe," says Bartels cautiously. "But we haven't proven that yet."

The idea that romantic love activates parts of the brain associated with addiction got Donatella Marazziti at Pisa University in Tuscany wondering if it might be related to obsessive compulsive disorder (OCD). Anyone who has ever been in love knows how consuming the feeling can be. You can think of nothing but your lover every waking moment. Some people with OCD have low levels of the brain chemical serotonin. Might love-obsessed people also have low serotonin levels? Sure enough, when Marazziti and her colleagues tested the blood of 20 students who were madly in love and 20 people with OCD, she found that both groups had low levels of a protein that shuttles serotonin between brain cells.

And what happens when the euphoria of "mad love" wears off? Marazziti tested the blood of a few of the lovers 12 to 18 months later and found that their serotonin levels had returned to normal. That doesn't doom a couple, of course, but it suggests a biological explanation for the evolution of relationships. In many cases, romantic love turns into compassionate love, thanks to oxytocin and vasopressin swirling inside the lovers' brains. This attachment is what keeps many couples together. But because attachment and romantic love involve different biochemical processes, attachment to one person does not suppress lust for another. "The problem is, they are not always well linked," says anthropologist Helen Fisher, who has written several books on love, sex and marriage.

IN THE WILD, about half of male voles wander the fields, never settling down with one partner. These "traveling salesmen," as Lowell Getz calls them, are always "trying to get with other females." Most females prefer to mate with their partners. But if they get the chance, some will mate with other males too. And, according to Jerry Wolff, a biologist at the University of Memphis, female voles sometimes "divorce" their partners. In the lab, he restricts three males at a time in separate but connected chambers and gives a female free range. The female has already paired with one of the males and is pregnant with his pups. Wolff says about a third of the females pick up their nesting materials and move in with a different fellow. Another third actually solicit and successfully mate with one or both of the other males, and the last third remain faithful.

Why are some voles fickle, others faithful? Vole brains differ from one creature to the next. Larry Young, a neuroscientist at Emory University, has found that some animals have more receptors for oxytocin and vasopressin than others. In a recent experiment, he injected a gene into male prairie voles that permanently upped the number of vasopressin receptors in their brains. The animals paired with females even though the two hadn't mated. "Normally they have to mate for at least 24 hours to establish a bond," he says. So the number of receptors can mean the difference between sticking around and skipping out after sex. Might these differences in brain wiring influence human faithfulness? "It's too soon to tell," Young says. But it's "definitely got us very curious."

How does evolution account for the often-conflicting experiences of love and lust, which have caused no small amount of destruction in human history? Fisher speculates that the neural systems of romantic love and attachment evolved for different reasons. Romantic love, she says, evolved to allow people to distinguish between potential mating partners and "to pursue these partners until insemination has occurred." Attachment, she says, "evolved to make you tolerate this individual long enough to raise a child." Pepper Schwartz agrees: "We're biologically wired to be socially monogamous, but it's not a good evolutionary tactic to be sexually monogamous. There need to be ways to keep reproduction going if your mate dies."

Many of our marriage customs, say sociologists, derive from the need to reconcile this tension. "As much as people love passion and romantic love," Schwartz adds, "most people also want to have the bonding sense of loyalty and friendship love as well." Marriage vows are a declaration about romantic love and binding attachment, but also about the role of rational thought and the primacy of mind and mores over impulses.

Scientists hope to do more than simply decode the biochemistry of the emotions associated with love and attachment. Some, like Insel, are searching for treatments for attachment disorders such as autism, as well as pathological behaviors like stalking and violent jealousy. It is not inconceivable that someday there might be sold an attachment drug, a monogamy pill; the mind reels at the marketing possibilities.

Lowell Getz, the grandfather of all this research, couldn't be more thrilled. "I spent almost $1 million of taxpayer money trying to figure out stuff like why sisters don't make it with their brothers," he says. "I don't want to go to my grave feeling like it was a waste."

Trevor Corson

Stalking the American Lobster

FROM *THE ATLANTIC MONTHLY*

Who knows better how to protect Maine lobsters from overfishing—scientists or the lobstermen themselves? A group of ecologists, armed with high-tech equipment and a healthy skepticism of conventional thinking, is coming up with a surprising answer, as the journalist Trevor Corson uncovers.

"Sir, I have a target, distance two hundred meters," the sonar operator said. "It looks big." The nuclear-powered submarine *NR-1* was hovering 600 feet underwater, on the edge of the continental shelf. Robert Steneck, a professor of marine sciences at the University of Maine, decided to check the target out. The helmsman nudged the sub forward, and Steneck, a short, energetic man with a thick red beard, slipped below the control room into the cramped observation module. There, through a six-inch-thick glass viewing portal, he was confronted with the biggest lobster he had ever seen. It was a female, about four feet long, weighing nearly forty pounds. She turned toward the sub as it came right up to her, nose to nose, and defiantly shook her claws.

Steneck is an unusual lobster scientist. Many of the leading scientists who track the North American lobster population do so mainly on computer screens in government laboratories, and from that vantage point lobsters appear to be in danger. From the mid-1940s to the mid-1980s Maine's lobstermen hauled in a remarkably consistent number of lobsters. But during the past fifteen years they have nearly tripled their catch, raising fears among many scientists about overfishing. The situation recalls the recent history of the cod fishery in New England, in which an exponential rise in the catch was followed by a devastating biological and economic collapse. In 1996, as lobster catches continued to hit all-time highs, a committee of the country's top government lobster scientists warned of disaster, and they have since recommended drastic management measures to save the fishery.

A failure in the lobster fishery—which has recently become the most valuable fishery in the northeastern United States—would be disastrous. Revenues from lobstering in 2000 topped $300 million. Nearly two thirds of the lobsters were caught in the waters off Maine, where some 4,000 fishermen earned $187 million at the dock for nearly 60 million pounds of lobster. And lobstering doesn't benefit only lobstermen: in Maine, for example, the fishery is a coastal economic engine that generates some $500 million a year altogether.

Most Maine lobstermen believe that their fishery is healthy, perhaps even too healthy. They worry not about a population collapse but about a market collapse. Even the lobstermen who admit that catches could decline don't see anything wrong with that. They say they're the lucky beneficiaries of a boom orchestrated by Mother Nature. If lobster catches soon return to more traditional levels, so be it.

The lobstermen argue that they are better biologists than the biologists are, and there's something to what they say. Fisheries scientists who gauge the effects of commercial lobster harvesting do so using techniques originally designed for tracking fish populations. Because fish are elusive and hard to study in the wild, estimates of how well their populations are faring rely heavily on mathematical models. But lobsters aren't fish. Many of them dwell in shallow coastal water and are easy to observe, though until recently few scientists had bothered to observe them. And unlike fish, lobsters aren't harmed by being caught. Baby lobsters, oversized lobsters, and egg-bearing lobsters that lobstermen trap and return to the sea are none the worse for having taken the bait—in fact, they've gotten a free lunch. Lobstermen know their resource more intimately than do many other kinds of fishermen, and they feel justified in telling the government that lobsters are doing well enough to be left alone. The trouble is that lobstermen tend not to have advanced degrees and scientific data to back up their claims, so their opinion carries little weight. But lately a new breed of lobster scientist has appeared along the Maine coast, epitomized by Robert Steneck on the *NR-1*. These scientists are ecologists, and they spend inordinate amounts of time underwater doing things almost no sane fisheries modeler with a computer and a comfortable office would ever do. They dig for days in the ocean floor to count tiny lobsters; they risk life and limb on shark-infested ledges seventy miles from shore to see how long lobster populations can survive predation. And they go lobstering with nuclear-powered submarines. Gradually they are concluding that some of the things lobstermen have been saying may be right.

BRUCE FERNALD HAS the ultimate lobsterman's physique: a low center of gravity and muscular shoulders. He has lived most of his life on an island called Islesford, off the coast of Maine. A pillar of the local community, he is often the one who gets a call when an elderly resident has a heart attack, and the one who rounds up the fishermen for a repair project on the public wharf. Fernald also takes a keen interest in lobster management and science. Along with several other Islesford lobstermen he has become one of Robert Steneck's most enthusiastic collaborators in the quest to collect data about lobsters. Steneck recognizes that lobstermen like Fernald and his colleagues spend far more time observing lobsters than he does, and that their knowledge can aid him in his research.

Fernald has always made his living by trapping lobsters across the 150 square miles of underwater boulder fields, gravel, and mud that surround the island. So have two of his brothers. He has never been down to see the terrain he fishes, but like a blind man who can read a face, he knows what it looks like—each gully, hillock, canyon, and plateau. His understanding of the lobster population around Islesford, developed during the course of a lifetime on the water, is similarly precise.

On a pitch-black morning in September 2001 the weather off Islesford was far from perfect: eddies and storm pulses were rolling in and battling with tidal currents inshore; the wind was picking up. Nevertheless, by 5:30 A.M. Fernald was on his boat, with his sternman. Cursing at a swarm of mosquitoes, Fernald checked the oil and cranked up the boat's 300-horsepower diesel engine. With the sky brightening, he pulled up to the thick mooring chain that tethered the boat to a two-ton slab of granite on the harbor floor, freed the vessel, and motored off. A hodgepodge of screens, instruments, and dials glared at him from the bulkhead and ceiling: engine readouts, bilge-pump alarms, a compass, a color Fathometer, a sixteen-mile-range radar, a GPS chart plotter. Also on the boat was a much less sophisticated bit of technology: a double-edged brass ruler known as the measure or the gauge. Since 1874 the measure has delineated the minimum size of a lobster that may legally be landed. In 1933 the State of Maine also instituted a maximum legal size. The main section of a lobster's armor, from the eye socket to the end of the back, is often referred to as the lobster's body but technically is called the carapace. In Maine the carapace must be no less than three and a quarter inches and no greater than five inches. Lobsters not meeting the measure are thrown overboard.

Out on the open water, Fernald gunned the boat to cruising speed while his sternman lifted the lid off the boat's bait bin, filling the cabin with the stench of herring. As the boat bounced against the chop, the sternman stuffed handfuls of gooey bait into small mesh bags with drawstrings. These he would soon be placing in the traps piled in the stern. Fernald had taken the traps up from shallow water a few days before and planned to drop them in deeper water this morning. He maintains 800 traps across a twenty-mile-long swath of ocean. He knows exactly where to place each one from one week to the next, March through December. He takes time off during the worst of the winter weather to repair his equipment.

During the summer Fernald keeps a third of his traps on short ropes near shore, strategically placed in certain coves and kelp beds, and near underwater boulders where he knows lobsters like to hide and hunt. In early September, though, lobsters begin to move offshore, so Fernald had already shifted much of his gear into middle-depth water—around a hundred feet. This morning's job was to set the first deepwater traps of the season. Seven miles out to sea Fernald pulled a dirty waterproof notebook from a tangle of electronic equipment and flipped through several pages of scrawled notes. He grabbed a pencil and jotted a few numbers directly onto the bulkhead next to his compass; then he squinted up at the GPS plotter above his head and keyed in a way point. He was headed for an underwater valley between Western and Eastern Muddy Reef. He was reassured to see his position confirmed by transmissions from four different satellites, but none of that was necessary: he could, if he had to, go back to navigating with nothing but landmarks and a magnetic compass, as his father still does.

Shortly, Fernald throttled down and studied the colorful blotches on his Fathometer screen, which was connected to a transducer on the bottom of the boat that bounced signals off the sea floor. The screen was painting the bottom as a thick black line at twenty-two fathoms, or 132 feet, which meant that Fernald was directly over the rocky ledge of Western Muddy Reef. He circled the boat a quarter turn and motored slowly east, watching the bottom on the Fathometer drop off and go from black to purple to orange, indicating a patch of cobble and then gravel where the ledge ended. Suddenly the line fell precipitously and settled into a mushy yellow haze at forty-seven fathoms, or 282 feet—a deep bottom of thick, dark mud. He was over the valley.

LIKE MOST LOBSTERMEN, Fernald believes that lobsters follow warmth. Fishermen think that many lobsters migrate in the spring toward land, to spend the summer in the sun-warmed waters near the shore, and migrate in the fall out to the mud in deeper water, far from the shallows that will soon be chilled by cold winds from Canada. The lobsterman must learn the lobsters' preferred routes along the bottom and intercept the animals on their pilgrimages. To succeed at his profession, Fernald therefore has to be an oceanographer, a sea-floor geologist, and a detective. Lobsters that migrate along the edge of an underwater canyon at one time of year may travel on the floor of the canyon at another time, so for Fernald to set his traps precisely can make all the difference.

When the lobsters show up near shore every summer, the first thing most of them do is go into hiding for a few weeks, to shed their old shells and grow larger ones. This process is called molting, and it is fraught with danger: not only must the lobster expose its jelly-soft body to the hungry world, but it may get stuck. The lobster's body shrinks, the old shell splits open, and the animal's twenty pairs of gills stop beating. The lobster has about an hour to wriggle free before it suffocates. The hardest part is pulling the large claw muscles through the narrow tracts of shell between them and the body—if the lobster can't do so, it will sacrifice one or both claws to live. Free of the old shell, the lobster gets its gills working again. Then for the next five hours it fills its shriveled body full of water. Artificially enlarged by liquid, the lobster then secretes the beginnings of a new shell, which will harden over the coming weeks. The new outfit should last a year or so, depending on the size of the lobster. The old shell is an excellent source of minerals, so the lobster eats some of it to quicken the hardening of the new one. What the lobster doesn't eat it buries with mouthfuls of pebbles, probably to hide the evidence of its weakness and also prevent rival lobsters from raiding its nutrient stash. While the lobsters are molting, Fernald takes advantages of the lull to haul his boat out of the water briefly for repairs and a new paint job.

By August "the shedders are coming on," as the lobstermen say, and the great autumn harvest begins. Dressed in their new shells, the lobsters are ravenous, and now millions of them meet the minimum carapace length for capture. Lobsters of this size enter the traps in droves. By the time the shedders begin to reach deeper water, Fernald must already have traps in place, which is why he was now setting a gauntlet of traps in the valley.

The task was complicated by the fact that the water had become a sloppy mess. A wave sloshed through an open panel in the boat's windshield and hit Fernald in the face and chest. He swore to himself, yanked the window shut, and shook himself off. Then he reached across the bulkhead and switched on the Clearview, a circular plate of glass in the boat's windshield that rotates eighty times a second—fast enough to fling off oncoming walls of water. "It would have been a lot easier to do this yesterday," he grumbled, "when it was flat-ass calm." He and his sternman pulled the first pair of traps from the pile in the stern and secured them on the rail so that they couldn't roll off before the buoy line was attached.

Fernald and his sternman arranged bulky coils of rope on the floor at their feet—carefully, because a tangle could cause mayhem. Fernald tied on a torpedo-shaped buoy, marked with his signature colors in Day-Glo paint; then he put the boat in gear and gave his sternman the signal to throw the first trap. It went over with a splash, and the workday was under way.

FERNALD AND his fellow fishermen on Islesford want to share their knowledge of lobsters, but few scientists have been interested in listening to them. With the arrival on the scene of Robert Steneck and other ecologists, however, that has begun to change. Steneck and others have spent long days at sea on the lobster boats of Bruce Fernald and his brothers, and have used the waters off Islesford as one of their research stations.

On a gorgeous morning in July 2001 Steneck was out in those waters, conducting a census of large lobsters a few miles from shore, the results of which might indicate that the lobster population is not in as much danger as some scientists think. Steneck's first task as an ecologist is to measure the abundance of lobsters and map their patterns of distribution. Baby lobsters and juvenile lobsters are relatively easy to study, because they live in shallow water; all Steneck needs to conduct his research on them is a scuba tank. But large lobsters are another matter—they've been known to live at depths exceeding 1,500 feet, though most of them probably don't venture much deeper than several hundred feet.

Steneck often explores the sea floor in a submarine, but on this trip he was using a submersible robot. The robot afforded him the luxury of staying above water, aboard the seventy-six-foot research vessel *Connecticut,* operated by the Marine Sciences and Technology Center of the University of Connecticut. The robot was a $160,000 piece of equipment known as a remotely operated vehicle, or ROV—an unmanned submarine that transmits video and other data from the ocean bottom to the mother ship through fiber-optic cables. The craft, operated with funding from the National Oceanic and Atmospheric Administration (NOAA), was called *Phantom III S2*, or, to the team of technicians accompanying it, just *P3S2*.

Also out on the water that morning, tending his traps in a forty-foot lobster boat, was Jack Merrill, an Islesford lobsterman who, like Bruce Fernald, has been in the business for nearly thirty years. Merrill is gruff, bearded, and thoughtful, and has dedicated much of his life to making lobstermen themselves the lobster's best advocate. To that end he, too, regularly collaborates with Steneck and other researchers. When Merrill caught sight of the *Connecticut* in the distance, he changed course and headed toward it. Twenty minutes later he throttled down and drew up under the *Connecticut*'s looming bow.

As Merrill pulled alongside, he was met by technicians carrying walkie-talkies and wearing orange flotation vests. Steneck emerged on deck, hailed Merrill, and pulled a notebook from his breast pocket. Merrill produced a notebook of his own and read off a few numbers to the scientist—numbers he would not have shared with his fellow lobstermen. This was one of his many small contributions to the quest for a better scientific understanding of lobsters. "That's where I've seen them," Merrill said. "Big ones, big time."

He then took the wheel of his boat and roared off across the sparkling water, back to his traps. Steneck climbed a steep stairway to the bridge, where he proceeded to map out the corrdinates Merrill had given him on a nautical chart. He nodded. "Two rock outcrops," he said. "Little underwater mountains. Just where you'd expect to find big lobsters."

Later in the morning, when the *Connecticut* was in position and Steneck was on his third cup of coffee, the ROV was put into the water. In the command module on the *Connecticut* the *P3S2*'s pilot, along with a copilot, Steneck, and one of Steneck's research assistants, monitored a bank of luminescent screens and instruments. The room echoed with sonar pings. Off to one side, with a video monitor of his own, sat the State of Maine's chief lobster biologist, Carl Wilson, a former student of Steneck's.

The pilot steered the ROV toward the bottom with a pair of joy sticks. On the video monitors a rain of plankton gave way to a lunar landscape of pebble fields and small boulders. *P3S2* was hovering at a depth of 104 feet. Its spotlights and three video cameras illuminated tall sea anemones growing on the rocks like stalks of broccoli. Fish darted around mussels, scallops, and the occasional starfish.

"This looks like a high-rent district," Steneck said. Steneck's research assistant switched on the video recorder and noted time and depth on a clipboard. Moments later a lobster antenna became visible.

"There's one," Steneck said. "He's hiding between those two boulders."

The pilot pressed his joy stick for a slow-motion dive. *P3S2* nudged the boulder, and the lobster's antenna twitched. The pilot pulled the ROV back, and the lobster emerged, strutting forward, claws extended and antennae whipping the water. If he had been able to see the ROV, the lobster might have been unnerved—but despite the fact that they are endowed with some 20,000 eye facets, lobsters have terrible vision. They have sensitive touch receptors, however, and an acute sense of smell. Two long antennae and thousands of tiny hairs on their claws and legs give them ample information about their environment. Like houseflies, lobsters can even taste with their feet. A second pair of shorter antennae, known as antennules, contain 400 chemoreceptors and give lobsters most of their hunting and socializing skills. But *P3S2* didn't emit a recognizable scent.

"That's it, baby," Steneck said to the lobster. "Work the camera." Steneck wanted a side view, in order to get a laser measurement. When the lobster turned to walk away, Steneck said, "Paint him with the lasers." A pair of laser beams hit the lobster squarely on a claw and the tail, providing a gauge of its size. This routine was more or less what Steneck and his team would be doing every day, ten hours a day, for the coming week.

"Is that another set of claws right there?" the ROV pilot asked, aiming for another boulder. "I don't think so," Steneck said. "That looks like a molt. Empty shell." But Steneck's attention was attracted by something else: the pebbly ground at the base of the boulder was a lighter color than the surrounding bottom, and had been carved into a small crater. "Hold it," Steneck said. "We've got recent sediment-reworking here. Let's take a closer look." The investigation paid off. The actual lobster, perhaps still soft from having recently shed its shell, was hiding around the corner, its presence betrayed by the burrow it had dug for itself.

The lobster wouldn't budge from its protected spot, but Carl Wilson saw a retreating shape in a corner of the screen. "Is that one?" he asked. The pilot changed course, and *P3S2* slowly gained on the lumbering lobster. This one clearly hadn't shed recently—large barnacles grew on its shell, an indication of its size, because bigger lobsters molt less often. Alerted to a presence behind it, the lobster spun, faced *P3S2* head on, lifted its claws wide, and ran directly at the ROV. "You're going to lose," the pilot said. At the last second the lobster seemed to reach the same conclusion, and it backed off.

THE FIRST LOBSTERLIKE DECAPODS probably evolved around 400 million years ago. Today there are thirty or so kinds of clawed lobsters, and forty-five species of clawless ones. By far the most abundant clawed lobster is *Homarus americanus*, or the American lobster. To the European explorers who arrived on the Maine coast in the 1600s, this greenish-brown crustacean looked familiar, because European waters are home to the American lobster's closest cousin: the bluish-black *Homarus gammarus*. But nowhere else in the world is *Homarus* as plentiful as it is in the waters off Maine. The explorers caught lobsters easily with long hooks or by dragging nets; later fishermen used a net hanging from an iron hoop and shaped like a cauldron—thus "pot," a term still used today to refer to a trap.

The basic design of the modern lobster trap was developed in the 1830s, and except for a switch from wood to wire, it hasn't changed much since. The number of traps in the water has changed dramatically, however. Records at the Maine Department of Marine Resources indicate that 50,000 to 100,000 traps were in use in 1880. Today some 2.8 million traps blanket the Maine coast.

A lobster trap is a wire-mesh rectangle almost four feet long, divided into sections: a "kitchen" and one or two "parlors." The bait bag hangs in the middle of the kitchen. On either side of the kitchen the wire is replaced by a ramp, made of knit twine, that ends in a small hole. Lobsters have an easy time walking up the ramp and through the hole into the kitchen; finding the hole and getting back out is more difficult. Many of those who can't find their way out are suckered into trying to escape on a third twine ramp—which leads to the parlor, designed to keep them stuck until the trap is hauled in by a lobsterman. Little lobsters have a Get Out of Jail Free card: the parlor is fitted with vents through which they can usually escape. Weather permitting, Bruce Fernald hauls his traps about every four days, and generally leaves them in the same area for several weeks. When lobsters begin to migrate elsewhere, he shifts the traps to follow them.

Today's lobster trap is a remarkably inefficient tool for catching lobsters. Winsor H. Watson III, a zoologist at the University of New Hampshire, and his graduate students have developed a device Watson calls a "lobster trap video," or LTV, which consists of a trap outfitted with a camera that looks down through a Plexiglas roof; a waterproof VCR unit; and a red lighting array for night vision. Watson can set the LTV on the bottom and run it for twenty-four hours to see how many lobsters enter the trap and what they do once they're inside.

Soon after a trap is set, lobsters smell the bait and approach. If the kitchen is unoccupied, more than half of those that approach will eventually enter and nibble at the bag of fish for about ten minutes. An astounding 94 percent of those walk right back out again. Furthermore, while one lobster is eating, other lobsters are often battling among themselves to be the next to enter, thus reducing the potential catch drastically—especially if the one eating also fights off any intruders on his meal. In one twelve-hour period recorded by Watson lobsters in the vicinity made 3,058 approaches to the LTV. Forty-five lobsters actually entered, and of those, twenty-three ambled out one of the kitchen entrances after eating. Twenty prolonged their stay by entering the parlor, but seventeen of those eventually escaped, leaving just five in the trap. Of those five, three were under the legal size. When Watson hauled the trap up, he'd caught a grand total of two salable lobsters.

Lobstermen like it that way. In Maine they have lobbied to outlaw other methods of catching lobsters, and during the past several years they themselves have imposed limits on the number of traps each lobsterman may set. Trapping provides a steady year-round job with time off in the winter, and it allows lobstermen to harvest only certain lobsters and throw back the undersized, oversized, and egg-bearing animals that are so crucial to the long-term health of the fishery. Most species that have collapsed from overfishing fell victim to radical improvements in fishing technology. "It's a very primitive trap we use," one lobsterman says, "and that's an important part of Maine law. As long as we keep using traps, we'll never catch them all. We're traditional in a lot of ways. I think that's going to save us in the long run."

The faster fishermen at Islesford can haul more than 450 traps in a single day. It's a demanding, manic routine, and it's dangerous. Most of the lobstermen on Islesford have tales of getting tangled in an outgoing rope as they race from one trap to the next; this can drag a man to his death in seconds. Two years ago a loop of outgoing line caught Jack Merrill around the ankle. He threw himself down as he was dragged aft and managed to lodge himself under the stern deck. His sternman rushed to the controls and threw the boat out of gear, saving Merrill's leg.

Fisheries scientists think that the hell-bent routine of lobstering is part of the reason lobsters are overfished; the race for profits, they feel, means that too many lobsters are getting trapped too soon. According to the scientists (though lobstermen dispute this), almost all of the annual catch now consists of new shedders—lobsters that have just molted up to the minimum legal size—instead of a more diverse sampling of sizes, and that doesn't bode well for the ability of the population to sustain itself.

JOSEF IDOINE IS employed by the National Marine Fisheries Service (NMFS), a division of the NOAA, as the chief federal biologist responsible for lobster. Idoine, who works at the NMFS laboratory in Woods Hole, Massachusetts, was not originally a lobster scientist. In college he majored in English literature, but the biological sciences and math had always captivated him. He later decided to pursue a degree in fisheries science. The professor with whom he studied modeled not only fish populations but insect ones as well, and Idoine realized that he could apply the same modeling techniques to lobsters. "Lobsters and insects both grow by molting," he says. "They're really not that different."

One problem Idoine faced—a problem that he continues to wrestle with today—is that scientists have yet to discover a reliable method for determining the age of a lobster. This means that although most fisheries scientists can rely on age data when they model fish populations, lobster modelers have to develop estimates of growth rates. Early in life lobsters molt frequently—up to twenty-five times in their first five years. After that they molt about once a year for a while, and when they're bigger, the rate drops again. Complicating the picture is the fact that female growth slows during reproduction, when energy goes into producing offspring instead.

In the 1980s Idoine and Michael Fogarty, a colleague at the NMFS, published papers that modeled a hypothetical lobster population. Modeling lobsters was in itself nothing new. Fogarty had already developed models describing the population dynamics of lobsters, and lobster scientists elsewhere had built careers around similar projects. But the Fogarty-Idoine model seemed to give scientists a better idea of how lobstermen might be affecting the lobster population's ability to sustain itself. The model suggested a common-sense idea: if lobstermen caught too many lobsters of too small a size, not enough lobsters would get the chance to grow larger, mate, and replace the lobsters being caught.

The Fogarty-Idoine model became an important part of a combined federal and state lobster-management system. Government scientists used the model to analyze the lobster population in the Gulf of Maine. The analyses led scientists to conclude that lobstermen were indeed risking the long-term sustainability of the resource by fishing too much. In the 1990s Idoine collaborated with another NMFS colleague, Paul Rago, to refine the model further; in its current version it is referred to as the Idoine-Rago model.

Lobstermen are suspicious of mathematical simulations like the Idoine-

Rago model. Jack Merrill, of Islesford, has long been one of the model's toughest critics. Like Idoine, Merrill studied both literature and science in college. When he started lobstering, in the early 1970s, he joined the Maine Lobstermen's Association (MLA) and soon became its vice president.

In the 1980s Merrill began collecting scientific papers on the lobster fishery. He noticed something strange: fisheries scientists had been using population models to predict the crash of lobster stocks for years, and so far not only had they been wrong but they'd had it completely backward—lobsters had done nothing but increase in numbers. When Fogarty and Idoine's papers came out, Merrill and other MLA officers met with Idoine and his colleagues at Woods Hole. "We asked, 'Why are you telling us we're overfishing?' " Merrill remembers. "They said, 'The formula tells us that you're overfishing.' "

The disagreement between Merrill and Idoine—and between almost all lobstermen and government scientists—boils down to a question of small lobsters versus big lobsters. Everyone agrees that in Maine's frigid waters only about 15 percent of lobsters are sexually mature at the minimum legal size. Lobstermen are harvesting prepubescents, which suggests to Idoine that very few female lobsters ever get the chance to mate. "That's what keeps me awake at night," Idoine says with a laugh. "Thinking about female lobsters." But the problem shouldn't be worth losing any sleep over, because a solution seems apparent. Government scientists have long recommended additional controls on lobster fishing, such as closed seasons and limits on the total number of traps in the water, but central to most management proposals has been raising the minimum legal size. That way more females would have a chance to mature and reproduce before they're caught. "Along with controls on fishing effort, raising the minimum size gives you a margin of safety," Idoine explains.

But Merrill and his colleagues in the MLA don't think Idoine's recommendations are necessary. They believe that the scientific models fail to factor in the margin of safety that lobstermen have built into their fishery for decades: a pool of large reproductive lobsters, protected not only by Maine's maximum-size restriction but also by a curious practice known as V-notching.

A CORNERSTONE OF Maine's conservation ethic, V-notching dates to 1917 and has been largely self-enforced by Maine lobstermen since the 1950s. V-notching is all about making babies. The sex life of lobsters does not get wide public attention, but it has attracted the interest of a small number of researchers. One of these is a biologist named Diane Cowan, a onetime professor at Bates College who is now the president of The Lobster Conservancy, a nonprofit research center dedicated to involving Maine coastal communities in lobster science.

Cowan once spent several months observing the behavior of a male lobster she had named M, which lived with one other male and five females in a tank at the Marine Biological Laboratory in Woods Hole, where Cowan later worked as a graduate student. Every night M would emerge from his shelter, boot all the other lobsters out of their shelters, and then return home. The females got the message: M was dominant. The females visited both of the males at their shelters, but M got far more lady callers than the other male. The visits were decorous at first: an interested female would insert her claws into the entrance of M's shelter and wiggle her chemoreceptor antennules to smell him. Then she'd urinate at him from the front of her head, releasing pheromones. In appreciation M would spread her urine throughout his apartment, by standing on tiptoe and fanning the water with his swimmerets—little fins along the bottom of the lobster's tail, arranged in five pairs.

Having ascertained mutual interest, the two abandoned all caution. M's primary concern seemed to be how soon the female would undress for him, and he would show his impatience by boxing the surfaces of her claws with the tips of his. Females can mate only after they shed their shells; Cowan thinks that M's boxing was a way of testing how hollow his lover's shell was in preparation for molting.

"One day I walked into the lab, and I thought there were three lobsters in M's shelter," Cowan says. It turned out to be not a ménage á trois but, rather, evidence of a conventional coupling. It was M, a female, and her molted shell. When a female that wants to mate is ready to molt, she lets the male know by placing her claws on top of his head, in a behavior scientists have termed "knighting." This apparently indicates to the male that he must protect her while she sheds her shell; scientists think the female may also release a sex pheromone that discourages the male from simply eating her, as he might under other circumstances. Once the female is undressed, the male gingerly lifts her soft body, flips her on her back, inserts a pair of rigid swimmerets into a pair of receptacles at the base of her abdomen, and passes his sperm into her. It's like the missionary position, but with double the genitalia.

Once the female's new shell had hardened, after a week or so, she moved out and a new female moved in. It turned out that all the females in the tank wanted to mate with M, and Cowan discovered that they were able to time their molts consecutively so that each would get a chance. Cowan describes the phenomenon as serial monogamy, and she has published papers on it; but she has also learned that females don't always follow its rules. "One night another female got in the shelter and took a chunk out of the [resident] female," Cowan says. "Lobsters get PMS—pre-molt syndrome. Before they molt, they have an activity peak and can go a little berserk." When Cowan altered the sex ratio in the tank, things got more confused. "When I had three males and just two females, the females couldn't make up their minds which male to stay with," she says. "They kept switching from male to male instead of pair bonding with just one guy. It was absolute chaos. It was horrible." Cowan altered the ratio further, and the results were even worse: "I tried to have five males and two females in the tank, but the males fought so aggressively that I had to take two of them out. Pretty soon even the remaining males had no legs left. They were walking around on their mouth parts because they were killing each other."

Lobstermen realize that producing offspring is a big commitment for a female lobster—up to twenty months of pregnancy and tens of thousands of eggs. At first the eggs develop inside her body, and she may wait for as long as a year after copulation to extrude them. Then she finds a secluded spot, rolls over onto her back, squirts the eggs onto the underside of her tail, and carries them around for another nine to eleven months. When they finally hatch and become larvae, she releases them into the ocean currents.

Once a female is carrying eggs, she becomes a kind of goddess to lobstermen. Most Maine lobstermen who find an "egger" in a trap will cut a quarter-inch V-shaped notch in her tail flipper (if she isn't notched already) before setting her free, and from then on it's illegal to sell her, whether she's carrying eggs or not. When she molts, the notch will become less distinct, but conscientious fishermen like Bruce Fernald and Jack Merrill always cut a new notch. "She's a proven breeder," Fernald says, "so we protect her." V-notched females and the oversized males that are protected by Maine's maximum-size law form a pool of reproductive lobsters called brood-stock. Lobstermen are convinced that brood-stock lobsters more than compensate for any deficiencies in egg production by smaller lobsters.

They're not necessarily wrong. A female lobster that has mated can extrude about 10,000 eggs if she has recently reached sexual maturity but ten times that if she is bigger. An older female lobster is also savvier: she can retain a male's sperm inside her body, perfectly preserved, for up to several years after copulation. She can use that sperm at will to produce a second batch of eggs. One Canadian researcher estimates that to achieve the egg production of a single five-pound female—a common size for a veteran V-notched lobster—more than ten smaller females would have to be protected. The lobstermen have a different way of saying the same thing: if V-notching were replaced by an increase in the minimum size, the increase would have to be so great that the only lobsters fishermen could legally catch would be too big to eat.

WHEN ROBERT STENECK STARTED scuba diving off the coast of Maine, in 1974, he was studying echinoderms and gastropod mollusks, but he kept getting distracted by lobsters. He'd been a researcher for the Smithsonian in the Caribbean, and he remains an internationally recognized expert on coralline algae, but in Maine he found a new calling. "There were all these lobsters and this huge industry that mattered to people," Steneck recalls. "I looked in the literature and realized that we knew almost nothing about lobsters in their natural habitats. I said to myself, 'Why the hell am I studying limpets?'" And there was a bonus in the study of lobsters: "At the end of the day you can have them for dinner."

At first Steneck's experiments were just for fun: he built underwater houses for the lobsters and found that the animals were partial to a section of plastic pipe with a rubber flap over one end. Soon Steneck was spending all his free time sawing up PVC pipe of different diameters, and the houses formed subdivisions. Given choices, the lobsters would shop around. It was almost as if they were picking out blue jeans: Steneck developed a record-keeping system that described smaller lobsters choosing "restricted fit" shelters and bigger ones going for "relaxed fit."

Steneck quickly became disenchanted with the suburban bliss he had created for his lobsters. He concluded that neighborly interaction was lacking and decided to change the zoning. But when he moved the shelters closer together, smaller lobsters moved out. "I set up video cameras," Steneck says, "and it turned out the dominant lobsters were fighting with their neighbors and evicting them." This wasn't as surprising as what Steneck discovered next. When he brought the shelters even closer together, the larger lobsters moved out, and smaller ones moved back in.

A complex set of hard-wired behaviors was govering the interactions that Steneck observed. Lobsters can be aggressive and cannibalistic, but most of the time they will dance delicately around one another to avoid unnecessary violence. If two lobsters of similar size are competing for a shelter, though, they may duel to establish dominance. Each opens its claws wide and spreads them threateningly. Circling like prizefighters, they urinate at each other and lash out with their antennae. If this behavior fails to scare away one of the lobsters, the next contest is a shoving match. Like bucks locking antlers, the lobsters put their outspread claws together and push in a test of strength.

When dominance has been established between two lobsters, if they meet again within seven days they won't bother dueling a second time. The dominant lobster will broadcast its status by secreting chemicals in its urine. Its swimmerets will pulse constantly to spread the urine into the surrounding water, and the losing lobster will recognize the scent and back off. Steneck's experiments revealed that even for a dominant lobster, though, avoiding fights in the first place was sometimes the best alternative. "If you're surrounded by a huge number of other lobsters, you have a choice," he says. "You can spend all day fighting, or you can move from an area of high population density to an area of low population density." Ironically, Steneck himself would soon be facing a similar choice.

Steneck was surrounded by fisheries scientists who thought about lobsters very differently from the way he did. He didn't agree with government modelers that lobsters were overfished. He also didn't think that encouraging more egg production would necessarily result in more lobsters. The complexities of the ecosystem in which lobsters live suggested to Steneck that any number of factors were affecting lobster abundance. The historical evidence from catches, and the large numbers of young lobsters Steneck saw underwater, suggested that the resource was healthy.

On Islesford, Jack Merrill also felt that the government's lobster experts were wrong. When he and his colleagues in the Maine Lobstermen's Association got wind of Steneck's research, they were intrigued. Ed Blackmore, the president of the MLA at the time, invited Steneck to a meeting of MLA officers. Steneck remembers it well. "I'd had very little contact with the industry at that point," he says, "and I didn't know anything—I showed up in a suit." To the room full of tough-skinned fishermen in boots and jeans, Steneck may have looked like just another scientist, but when he started talking, they sat up in their chairs. "We didn't agree with everything he said," Merrill recalls, "but it was the first time we'd heard a scientist say anything that made any sense."

By the summer of 1989 Steneck's claim that lobsters weren't being overfished had provoked the ire of government scientists. Steneck was summoned before a committee of two dozen lobster experts and interrogated. "At academic conferences my work had always been well received," he says. "Everything I said to that committee was later published in international journals. But with the committee it was dead on arrival."

The committee declared Steneck's findings irrelevant, but it was too late.

Lobstermen in both the United States and Canada had gotten wind of Steneck's research, and many of them embraced his ideas. As a result, support for a four-stage increase in the minimum size, which lobstermen had agreed to a few years before, evaporated. Negotiations involving President George Bush and Canada's Prime Minister, Brian Mulroney, failed to save the deal. The first two increments of the increase had already been enacted, but state legislatures blocked the remainder. Leading government scientists blamed Steneck.

STENECK BEGAN to collaborate with two ecologists—Lewis Incze and Richard Wahle, both with the Bigelow Laboratory for Ocean Sciences, in Boothbay Harbor, Maine—to design an alternative to the government's modeling system. To gauge the health of the lobster population, the team developed a system for measuring the abundance of lobster larvae in the water, baby lobsters on shallow bottom, and brood-stock lobsters in deep water.

In the mid-1980s Wahle had realized that there was a huge blind spot in the understanding of the lobster's life cycle. "Lobster larvae just disappeared into a black box," he recalls, "and came out four or five years later, when lobsters showed up in fishermen's traps." He started asking around. From a fellow diver and lobster researcher he heard about an area where an unusual number of small lobsters had been seen. The information was old, but Wahle went out with his scuba gear and had a look anyway. "It was unbelievable," he remembers. "You'd turn over rocks and boulders, and every little rock had a tiny lobster under it. It was clear that we'd hit on something. This was a nursery ground."

Wahle developed a kind of giant underwater vacuum cleaner for collecting baby lobsters, counting them, and returning them to the bottom unharmed. After sampling several locations in this way, he concluded that shallow coves with lots of cobbles were the ideal habitat for baby lobsters. But not all such coves actually were nurseries. To determine why, Wahle teamed up with Incze, whose specialty was catching larvae by towing fine nets behind a research boat—a job that one of his colleagues has likened in difficulty to collecting insects with a butterfly net towed behind a helicopter. With Incze trawling for larvae on the surface and Wahle diving below, the two men discovered what distinguished the best nursery grounds: they had the best conditions for the delivery of larvae. "The hot spots are where you get a convergence of good habitat, ocean current, and prevailing winds," Wahle explains. The currents and winds were delivering larvae to the nursery grounds from wherever they had hatched.

Surprisingly, lobster larvae themselves have a small say in where they end up. Most fish larvae are helpless creatures, utterly at the mercy of water flow, and so are lobster larvae, up to a point. After hatching they progress through three planktonic stages, a process that probably takes several weeks, during which they are carried by whatever prevailing current catches them. Then they molt to a fourth stage, which biologists call post-larval—or, more affectionately, the "super-lobster" stage, because like miniature Supermen, these little lobsters fly through the water with their claws outstretched. "They can swim around powerfully in a horizontal direction, but it's more the vertical dives they're able to control," Wahle explains. This is the only period in a lobster's life when it can swim forward, propelling itself with the swimmerets under its tail, and it lasts no more than a week or so. "They have a biological clock ticking," Wahle says, so the search for protective cobbles is frantic. If a superlobster dives and can't find a nice spot to settle down, it launches itself back into the current and tries again.

Once Wahle had mastered techniques for counting baby lobsters in their nurseries, he and Steneck realized they might be on the verge of a new era in lobster management. They developed a rigorous sampling protocol, trained a team of student divers from the University of Maine, and, in 1989, initiated a series of annual measurements that they thought might work as a predictive system: a future increase or decrease in lobster catches ought to show up in advance as a fluctuation in the number of larvae and young lobsters the ecologists observed.

Meanwhile, the ecologists began to monitor lobster brood-stock as well. Lobstermen like Bruce Fernald and Jack Merrill had a great deal of data to offer the scientists—fishermen hauled up egged lobsters in their traps every day. Steneck put Carl Wilson, his graduate student at the time, in charge of an ambitious effort to get university interns out on lobster boats counting lobsters, especially V-notched lobsters—a technique called sea sampling. At first the sampling trips took place during the summer months and weren't especially productive. "Jack kept harping on Bob," Wilson recalls, "saying that the fall was when the egged lobsters really migrated offshore—that's when we should do it." In the autumn of 1997 Wilson decided to go to Islesford, spend a day on Merrill's boat, and see for himself. "Jack was right," he says. "It was just mind-boggling. Huge eggers, huge V-notchers, all these egg-bearing females that you never see during the rest of the year, came up in the traps."

While his students worked aboard lobster boats, Steneck descended to the sea floor in a submarine to quantify brood-stock lobsters, including those too large to come up in traps. Steneck continues these dives today, using ROVs, and

Wilson, as the State of Maine's chief lobster biologist, now heads an expanded sea-sampling program. So far the data appear to support the lobstermen's contention that the population of large reproductive lobsters is actually bigger than mathematical models suggest.

But the data the ecologists have collected on lobster larvae and baby lobsters complicate the picture. Through 1994 they observed an abundance of both larvae and babies on the bottom. As those baby lobsters grew to marketable size, lobster catches went even higher, hitting records in 1999 and 2000. In 1995, however, Incze had seen a sudden drop in larvae, and Wahle had seen a widespread decline in the baby lobsters he was counting on the bottom. The two teamed up with Michael Fogarty, the population modeler at the National Marine Fisheries Service, to develop a different kind of model—one that would use the ecologists' new data to predict future catches. After five years of low counts of larvae and babies, in the fall of 2000 Incze, Wahle, and Fogarty announced their findings at a scientific conference. Steneck announced that his scuba surveys of juvenile lobsters also reflected a downward trend in some areas.

The ecologists' system had yet to be proved predictive, but Steneck, Wahle, and Incze decided to issue a press release announcing that they had witnessed a decline in larval and baby lobsters. They admitted that the implications for lobstermen were still unclear, but they hazarded a guess that the stunningly high numbers of legal-sized lobsters that fishermen had been catching might start to drop along parts of the coast that coming fall. It was the first time a statement from Steneck had sounded anything like what government scientists had been claiming for decades.

But what he was saying was quite different. Government scientists had been arguing that low egg production could lead to a population collapse. Steneck and his colleagues were suggesting that the decline they had witnessed had nothing to do with either egg production or a collapse. The ecologists believed that egg production was reasonably stable—perhaps even at a surplus, thanks in part to V-notching—but that something was affecting how many eggs survived. Very probably, they thought, fewer larvae than before were getting to the nursery grounds. The question was why.

STENECK, WAHLE, AND INCZE are convinced that the culprit is the ocean itself. They don't deny that a certain number of eggs is necessary for a sustainable fishery, but they argue that even if lobstermen were to protect many more lobsters and allow them to produce eggs, larval biology and ocean currents would still have the final say in whether those eggs became new lob-sters. And, as any fisherman knows, the sea is fickle.

Ocean currents are decisive because they carry the larvae from the place where their mother hatches them to the nursery grounds where they settle on the bottom. Lobsters bearing eggs are found in many areas along the coast of Maine, but Steneck has identified what appear to be special concentrations of them in certain places. These findings indicate that some larvae are arriving at their nursery grounds having been hatched just around the corner, but others are probably coming from hundreds of miles away. Now that the ecologists have an idea where egg-bearing lobsters are on the one hand and where nurs-ery grounds are on the other, they hope to track the exact oceanographic links between the two.

Incze, the oceanographer of the group, speaks of "retention" of local larvae and "delivery" of distant larvae. He believes that depending on currents, any given nursery ground can experience both, or one without the other, or—in the worst case—neither. "As oceanographic conditions change and steer cur-rents toward shore, bringing larvae with them, you can have a large increase in the number of settlers coming from different areas of egg production," Incze says. "Alternatively, when those currents steer the water offshore, you can have an entire region decline in settlement." That might explain why the number of larvae and baby lobsters decreased in the second half of the 1990s.

Incze is studying why these changes in oceanographic conditions occur. "Ice melting in the Arctic, cloud cover, and prevailing winds can all affect how water moves around the Gulf of Maine in specific ways," he says. Incze will also be examining the possible influence of the North Atlantic Oscillation, a varia-tion in the distribution of high- and low-pressure systems over the Atlantic Ocean that affects weather patterns and ocean currents. "The North Pacific Os-cillation has been well studied," Steneck points out, "and we know it affects the Hawaiian spiny lobster." If the ecologists can learn why currents vary, and can track these variations within the Gulf of Maine, their understanding of what drives the size of the lobster population from one year to the next will dramat-ically improve.

Scientists will never be able to prevent declines in the lobster population caused by ocean currents, but Steneck and his colleagues hope to be able to warn lobstermen of them accurately, by monitoring the abundance of larvae and baby lobsters. And if Josef Idoine turns out to be right, and egg production drops off dangerously, that should show up too, as a decrease in the number of big lobsters being counted by sea sampling and submarine dives. Either way, the catches of the coming decade will reveal whether the system the ecologists have developed is indeed predictive. Steneck is confident it will work. It's not a ridiculous notion: in recent years a similar system has successfully predicted catches in the rock-lobster fishery of western Australia.

WHETHER OR NOT lobsters are being overfished, lobstermen face some serious problems. If the banner years end and catches return to their previous levels, overfishing might become a more plausible danger, because there are far more traps in the water than there used to be. And even if no biological disaster ever occurs, an economic one might. Fishermen who have invested too heavily in their equipment will suffer if catches decline, as will families who have grown accustomed to a higher standard of living. Some lobstermen fishing today have no memory of the slower-paced, less lucrative kind of lobstering that the older generation knew. That is because they started lobstering recently—after the collapse of other fisheries. On the whole, however, lobstermen in Maine are thoughtful and broad-minded stewards of a communal resource, and they understand that fishing sustainably is in their best interest. As one Islesford lobsterman puts it simply, "We throw back for tomorrow."

Bruce Fernald's father, Warren, is confident that the lobster population is in good shape. But given what he's seen in his half century of lobstering, he also admits that a decline in catches might be just what the industry needs. "I always relish a shakeout," he says. "Sometimes scoundrels get into the fishery. After a shakeout they don't do so well. The guys that have been hanging in there do okay."

Halfway through the summer of 2001 Bruce Fernald was afraid a shakeout might already have arrived, too soon for his taste. The lobstering in the spring and summer had been mysteriously dismal, and some of the Islesford lobstermen were beginning to worry that the annual run of shedders would never materialize.

On the July 2001 day when Robert Steneck was exploring with the ROV in the waters off Islesford, Fernald finished hauling his traps early—there wasn't much in them—and decided to swing by the *Connecticut*. The afternoon breeze was whipping up a light chop, so Fernald had to jockey his boat up to the side of the *Connecticut* with agile flicks of the throttle and well-timed twirls of the wheel. Steneck emerged on deck and traded banter with Fernald across the trough of seawater splashing between them.

"So far this is the worst season I've ever had!" Fernald shouted over the thump of his diesel engine. "But I'm seeing more oversized lobsters, V-notched lobsters, and eggers than ever."

"That's good!" Steneck shouted back. "We're picking up a lot of larvae in the water. For the long term, maybe things aren't so bad. And I think you're going to start seeing shedders in your traps any day now. We're seeing them on the bottom."

"I'll believe it when I see it," Fernald answered. He backed his boat away from the research ship, leaving a frothy wake. "Come by the island for a beer sometime!" Fernald shouted, saluting Steneck as he pulled away.

A few minutes later the VHF radio on Fernald's boat sputtered to life, and a scratchy voice came over the airwaves. It was Jack Merrill, who had earlier in the day given Steneck coordinates for finding big lobsters. He was calling Steneck on the *Connecticut* from where he was fishing, fifteen miles out to sea.

Steneck responded, and after a brief exchange said to Merrill, "I don't know if it makes any difference to you where you're fishing, Jack, but I just told Bruce that over here we're starting to see some shedders."

"Oh, yeah?" Merrill said, sounding incredulous. "Throw a few in my traps, will you?"

"Yeah, right!" Steneck said.

The voice of another Islesford fisherman crackled through on the radio. "You saw shedders?" he said, his tone almost pleading. "Where the hell are you? Stay right there, I'm on my way."

As it turned out, the lobstermen of Islesford didn't need submarines to find their lobsters after all. In the middle of August the shedders came on like never before. The fishermen counted their blessings and fished like crazy.

SIDDHARTHA MUKHERJEE

Fighting Chance

FROM *THE NEW REPUBLIC*

What is the best way for the government to fund scientific research: by directing scientists toward a particular goal or simply letting them decide what areas to study? Taking a close look at the question, medical researcher Siddhartha Mukherjee cannot deny that both approaches have brought success. As he reports, however, the path one scientist took to the development of a potentially effective antidote to the anthrax bacillus may show the best way.

In the summer of 1987, nearly 15 years before words like "anthrax" and "bioterror" saturated our vocabulary, an unassuming biology professor named John Collier went to hear a graduate student give a talk on *Bacillus anthracis.* Collier was interested in anthrax because of a peculiar property of the bacteria. Anthrax, like a few other microbes, extrudes a deadly toxin that is capable of wreaking havoc on human cells. In fact, anthrax's most frightful symptoms—the coal-colored dimples that erupt into pustules and the volcanic hemorrhages that pour out of the organs—are actually just manifestations of this toxin's actions. And yet, despite decades of intensive military research on anthrax, little was known about how the toxin worked or how humans could be protected from its effects.

Collier was actually something of an aficionado of bacterial toxins. In the mid-1960s and 1970s, he and a group of colleagues had deciphered the mechanism for the toxin made by diphtheria. In his laboratory at Harvard, Collier had another set of graduate students tinkering with a toxin produced by a bacterium called *Pseudomonas aeruginosa*. So while listening to that talk in 1987, Collier decided that anthrax would be his next target. "It wasn't even a conscious choice," he recalls. "As a biologist, you just *had* to be intrigued."

In the wake of September 11, Collier's reasoning is worth remembering. Since the World Trade Center fell, there have been numerous exhortations to the nation's scientists to turn their attention to the terrorist threat. The Pentagon and the Department of Health and Human Services have issued calls to researchers for proposals that pertain to the War on Terrorism. Nearly $2 billion in funding is now available for anti-terror research. Georgia Senator Max Cleland has even called for a new anti-terror Manhattan Project to be led by the Centers for Disease Control (CDC), declaring: "This is a race for the best minds, the best talent, and the best technology we can find in the realm of biological, chemical, and radiological warfare."

But Collier didn't embark on a 15-year investigation into anthrax because he was worried about bioterrorism or germ warfare. In fact, his early scientific papers and grant applications don't even mention the words. Collier spent thousands of hours picking the toxin apart, piece by piece, simply because he was curious about the basic biology and chemistry of the proteins. And in retrospect, there was probably something inherent in that curiosity—in Collier's becoming "intrigued" with anthrax toxin as a quandary of basic biology—that ultimately accounted for his success.

All of which suggests a paradox. In the post–September 11 world, it's tempting to think of curiosity-driven science as an anachronistic luxury. Wars inevitably make nations pragmatic about spending. And so there is already public pressure to funnel billions of dollars into applied research, into research that directly intersects with the dramatic changes in the political sphere. But Collier's story suggests the pitfalls of such an approach: Ironically, Collier may have cracked the mystery of anthrax toxin precisely because he *wasn't* out to curb the threat of bioterrorism. In other words, even in these pressured times, we may be better off leaving such scientists alone—to follow their curiosities wherever they lead.

THIS DICHOTOMY—between science driven by curiosity and science driven by applications—began long before the War on Terrorism. In the immediate aftermath of World War II, Franklin Roosevelt found himself embroiled in a similar national debate. Back then, the case against funding curiosity-driven research seemed obvious. For much of the American public, the Manhattan Project—conducted by scientific SWAT teams in military laboratories—had shown that applied science was far more efficient than the arcane musings of academic namby-pambies tucked away in university labs. On August 7, 1945—the morning after the Hiroshima bombing—*The New York Times* declared, "University professors who are opposed to organizing, planning and directing research after the manner of industrial laboratories . . . have something to think about now. A most important piece of research was conducted on behalf of the Army in precisely the means adopted in industrial laboratories. End result: an invention was given to the world in three years, which it would have taken perhaps half-a-century to develop if we had to rely on prima-donna research scientists who work alone . . . a problem was stated, it was solved by teamwork, by planning, by competent direction, and not by the mere desire to satisfy curiosity."

If the trends augured by that editorial had persisted, Collier would have never even *started* work on anthrax toxin. Anthrax—like the Bomb—was, after all, a military problem. In fact, by the mid-'60s, anthrax research was already in full swing in the U.S. Army laboratories at Fort Detrick, Maryland. Scroll back through the early scientific literature on anthrax, and you'll find scores of scientists from Fort Detrick plugging away on various aspects of the microbe. They produced microscopic studies on how the organism forms spores, and careful disquisitions on the structure of its outer coat. In 1967 a journal called *Federation Proceedings* ran a whole seminar series on anthrax, most of which emerged from labs at Fort Detrick.

Superficially, these studies were flawless. But it's impossible to compare them to the rigorous brilliance of Collier's research—to the carefully dissected experiments, or the complete immersion in the biology of the toxin that characterized Collier's early years at Harvard. Even Collier, who has a reputation for reticence, agrees. "The military researchers contributed greatly to anthrax research," he says, "but the mechanism of the toxin wouldn't have been so quickly solved had university labs not gotten involved as well."

Indeed, in Collier's hands, the search for an anthrax antidote took a completely different turn. For ten years, beginning in 1987, Collier and his team delved deeper and deeper into the basic biology of anthrax toxin. For Collier, anthrax toxin became a sort of intricate wind-up toy, whose inner clockwork could be solved only by taking the whole unit apart. The toxin itself, it turned out, was actually a conglomerate of three distinct proteins: Lethal Factor, Edema Factor, and Protective Antigen. Each member of this trio seemed to play a critical role in the toxin's action. Protective Antigen led the charge: A fragment of the protein bound itself to the surface of cells and formed aggregates on the cell surface. Lethal Factor and Edema Factor then bound to these aggregates, entered human cells through a "pore" created by the Protective Antigen fragments, and proceeded to poison the cell.

By the late 1990s Collier's work on these details began to yield astonishing payoffs. Once Protective Antigen's critical role had become clear, Collier realized that he might be able to thwart the toxin by blocking Protective Antigen directly. And last year Collier's team published two landmark papers describing not just one but two such anti-toxins directed against Protective Antigen. Preliminary studies with the new drugs far exceeded Collier's modest expectations. Laboratory animals medicated with either of the molecules became totally immune to lethal doses of anthrax toxin. If the same sort of drugs worked in humans, Collier argued, they could potentially combat even the "late stage" of anthrax, the frightful crescendo of the illness, when the disease can no longer be curbed by conventional antibiotics or vaccines.

In short, anthrax turned the logic of the Manhattan Project on its head. In the brief span of 50 years, the paradigm of research had dramatically changed, with academic scientists—the tweedy "curiosity-driven" professors once mocked by the *Times*—playing a critical role in understanding the toxin and the military researchers lagging behind.

WHAT HAPPENED? The answer lies in the complete overhaul of science funding that began during Roosevelt's presidency. In 1944 FDR asked the head of the Office of Scientific Research and Development, an MIT-trained engineer named Vannevar Bush, to devise a science plan for postwar America. And, bucking prevailing sentiment, Bush produced a manifesto, entitled *Science: The Endless Frontier*, that would transform the compact between science and society. Bush's plan rested on one key assumption: that, in the short term, people would never grasp the true value of basic science. If basic and applied science were allowed to mix and compete freely, the latter would inevitably drive out the former. The only way basic science could survive—something Bush wanted to ensure—would be to completely insulate it from that competition, leaving basic scientists to pursue their work in peace.

The institutions charged with protecting basic science were the National Institutes of Health (NIH) and the National Science Foundation (NSF)—near-autonomous scientific foundations that grew out of the Bush plan in the 1950s. By that decade's end, both institutions had become independent science-funding bodies, run for and by scientists. Shielded from political accountability, curiosity-driven scientists were given plush Ivy-league sandboxes and growing public largesse to spend. They could dismantle the structures of recondite proteins if they so pleased, or sequence the genomes of fruit flies—as long as other scientists deemed their goals scientifically worthwhile.

Princeton political scientist Donald Stokes compared Bush's compact to a "deal" between society and scientists. Society would invest unflinchingly in the basic sciences without insisting on premature technological rewards. Politicians and bureaucrats wouldn't go up to someone like Collier and say, "Now wait a second, haven't we funded research on anthrax toxin for five years? Where's the antidote?" In return, scientists would pursue basic research in good faith—being broadly receptive to the technological innovations that might emerge. It was this deal that allowed Collier to sit for 15 years at Harvard, plumly funded by NIH grants, calmly chipping away at the structure of a bacterial toxin, before finally producing a new drug that would block it.

BUT HOW DO WE KNOW THAT "programmatic" research—research that aims to find cures for applied problems within specified periods of time—wouldn't benefit society more? In a provocative article in *Nature Medicine* in 1997, the physician and scientist Richard Wurtman argued that it almost certainly does. His case study was AIDS. AIDS, remember, was identified as a clinical entity among gay men in San Francisco and New York just 17 years ago. In 1993, at the heyday of the epidemic, scientists were uniformly pessimistic about whether HIV would ever be a curable infection. And yet, a mere eight years later, triple therapy with antiretroviral medicines—for those who can afford it—has made HIV into a largely treatable disease. That turnaround time of 17 years—between the discovery of AIDS and the discovery of a therapy—is a truly astonishing accomplishment in the history of medicine.

Wurtman contends that this rapid pace of drug discovery was the result of a paradigm shift in HIV research. In the early '90s, AIDS activists began campaigning ferociously for federal dollars to combat the disease. But instead of picketing the NIH for "more research," they demanded "effective treatments"—i.e., they didn't demand greater inquiry; they demanded better results. As Wurtman tells the story, scientists responded to these demands by revising their own ideas about research. Instead of digging in their heels against "mission-oriented," or programmatic, research, they began actively scouting for antiviral therapies—even before much of the basic biology of the virus was fully understood. Prodded along by impatient activists, HIV researchers somehow picked up on what Wurtman calls an "implicit call for accountability." And that combination of public accountability and programmatic focus, Wurtman believes, brought about the effective anti-AIDS drugs in record time.

But programmatic research has problems that Wurtman's parable doesn't acknowledge. The first is counterintuitive: It may be more expensive. In 1999 the NIH ran a complex accounting project in which it attempted to correlate the amount of money spent on a particular disease with some measure of the actual years of life that the disease had claimed from Americans (the so-called disease burden). The goal was to determine whether federal science money was being spent in a disease-proportional manner.

The results were astounding. For most diseases—even money guzzlers like cancer and cardiovascular research—federal spending was more or less proportional to the disease burden. The big exception was AIDS. In 1996 the NIH spent proportionally more public money on AIDS than on any other disease. Crudely put, programmatic research had indeed produced remarkable anti-AIDS drugs, but at an enormous price. And it's not hard to understand why the cost was so high. Once you declare "war" on a disease, Collier argues, "you would get plainly bad science—a lot of junk aimed at getting some of that pork-barrel money." After all, you have taken the ultimate funding decisions away from scientists and given them to politicians.

BUT THE PROBLEM with programmatic research isn't only that it may be more expensive; it's that it leaves little room for a critical feature of the discovery process: serendipity. Indeed, the history of HIV research itself offers eloquent testimony to the role of serendipity in science. Anti-HIV therapy was revolutionized in the mid-'90s with the discovery of a novel class of antiretroviral chemicals: protease inhibitors. These inhibitors block a critical step in the viral life cycle, the point when the virus is just about to launch itself out of an infected cell into another. That process, it turns out, is mediated by a critical enzyme called HIV protease.

HIV protease closely resembles another such protease found in the human kidney—renin—which is involved in regulating blood pressure. In fact, researchers had been scouring for an inhibitor for renin long before HIV arrived on the scene. These kidney scientists and blood-vessel biologists—some in pharmaceutical companies and some in academic laboratories—had nothing to do with the much-publicized "War on AIDS." But when the search for protease inhibitors intensified in 1990, it was this prior work that HIV virologists used to suddenly make the critical connection between the two completely unrelated fields—jump-starting the search for these revolutionary new drugs.

EXAMPLES OF such serendipitous breakthroughs abound in the folklore of science. Sylvia Wrobel, writing in the scientific magazine *FASEB Journal,* relates the story of a mysterious and lethal infection that broke out in New Mexico, Arizona, Colorado, and Utah in 1993. Within days of the outbreak, the CDC deployed a team of top-notch scientists to ferret out the cause of the infection. Epidemiologists, virologists, and molecular biologists swarmed the desert looking for clues to its source. But the observation that clinched the discovery came from an extraordinarily unlikely source. Ecologists at the University of New Mexico had been tracking the population of deer mice to collect data for a completely unrelated project. And looking back at the ecologists' data, the CDC scientists noticed that the human infections seemed to occur just when the population of mice in the area swelled. With that clue in place, it took just a few weeks to discover a novel virus—called hantavirus—from the mice. The hantavirus discovery has long been considered a landmark CDC achievement. And yet, had the CDC declared a "War on Southwestern Fever," it's hard to imagine it would have funded a project on deer-mouse population ecology as part of that effort.

The point is that scientific discoveries often happen when they are least expected. Disparate nodes in knowledge are inexplicably connected through secret passages. And the danger is that a post–September 11 focus on programmatic research might demolish this *Looking-Glass* universe. One of the everlasting quirks of curiosity-driven science is that it allows kidney biologists to find themselves at the forefront of HIV research or mouse ecologists to become hantavirus hunters. And perversely, the more narrowly you define a scientific goal—hoping to focus and streamline discovery—the more you potentially logjam the discovery process itself, setting technology back as a result.

But for John Collier, there's a final argument for curiosity-driven research, and it has little to do with science money or scientific serendipity. In fact, it lies outside the reach of science itself—in the foggy realm of personality. And Collier could only explain it by walking me through a critical fork in his own life.

In 1959 Collier, like any other ambitious college graduate curious about science, had to choose between going to medical school and becoming a basic biologist. He spent an entire year mulling the choice. At the end of his final summer in college, he chose to spend an extra year working in a laboratory. It would be "just for a year," he told himself—and then he would consider medical school again.

Collier's year in that laboratory wasn't particularly memorable. There wasn't any single moment of scientific epiphany. But something ineffable happened to him—a transformation he struggles to describe. When application season came around again, he said, medical school was no longer an option. Maybe it was "merely inertia," but he had become, as he put it, "somehow drawn in."

Collier told me all this in his office, a sparse, sunny room on the fourth floor of the "Quad" of buildings that houses Harvard Medical School. The office is stockpiled with scientific journals. And just across from his desk, Collier has hung an enormous poster of the structure of diphtheria toxin, magnified more than a million times, the way a child might put up a poster of a basketball star. There is an empty space beside it, he explains, where a similar blowup of *Bacillus anthracis* toxin will one day hang.

People like Collier—people who frame bacterial toxins in mega-size posters on their walls—may never be "drawn in" by programmatic research. They can't be recruited for an applied research project—a "War on Anthrax," for instance. For Collier, and hundreds of eccentrics like him strewn across academic campuses throughout the United States, curiosity-driven research isn't just the best way to do research; *it's the only way.*

In retrospect, America's fateful decision to provide massive funding for basic science is deeply ironic. Vannevar Bush's model was adopted—and continues to be accepted—with scarcely any empirical evidence to support it. Expert committees weren't appointed to approve it; critics weren't called in to drum up statistics to illustrate its shortcomings. The everlasting paradox of American science is that it is based on rules that wouldn't survive even the most rudimentary scientific scrutiny.

But nonetheless, the history of anthrax research suggests that the technology transfer Bush dreamed of has, indeed, come to pass. This summer, when Collier and his co-workers found a way to block anthrax toxin with a novel drug, a group of scientists got together to seed a new pharmaceutical company based on their discovery. That company will presumably run clinical trials on the anthrax antidote—perhaps in coordination with medical researchers—in the hope that the FDA will eventually approve it for human use.

Collier is relieved about this transfer of responsibilities. Drug development isn't his passion; he wants to return to the laboratory bench, to further explore the basic structure of anthrax. He has to renew his grant from the NIH. He has graduate students and postdoctoral researchers to mentor. But as far as finding a potential antidote to anthrax goes, a scientific cycle may be coming to a close. And John Collier—who never imagined himself a poster child for curiosity-driven science—seems deeply satisfied with that.

MICHAEL KLESIUS

The Big Bloom

FROM *NATIONAL GEOGRAPHIC*

> *Flowers beautify, symbolize romance, fuel obsessions. For all their ubiquity in our lives, scientists still are not entirely sure how they originated. (Darwin called this "an abominable mystery.") Michael Klesius uproots the 130-million-year history of the flowering plant.*

In the summer of 1973 sunflowers appeared in my father's vegetable garden. They seemed to sprout overnight in a few rows he had lent that year to new neighbors from California. Only six years old at the time, I was at first put off by these garish plants. Such strange and vibrant flowers seemed out of place among the respectable beans, peppers, spinach, and other vegetables we had always grown. Gradually, however, the brilliance of the sunflowers won me over. Their fiery halos relieved the green monotone that by late summer ruled the garden. I marveled at birds that clung upside down to the shaggy, gold disks, wings fluttering, looting the seeds. Sunflowers defined flowers for me that summer and changed my view of the world.

Flowers have a way of doing that. They began changing the way the world looked almost as soon as they appeared on Earth about 130 million years ago, during the Cretaceous period. That's relatively recent in geologic time: If all Earth's history were compressed into an hour, flowering plants would exist for only the last 90 seconds. But once they took firm root about 100 million years ago, they swiftly diversified in an explosion of varieties that established most of the flowering plant families of the modern world.

Today flowering plant species outnumber by twenty to one those of ferns and cone-bearing trees, or conifers, which had thrived for 200 million years before the first bloom appeared. As a food source flowering plants provide us and the rest of the animal world with the nourishment that is fundamental to our existence. In the words of Walter Judd, a botanist at the University of Florida, "If it weren't for flowering plants, we humans wouldn't be here." From oaks and palms to wildflowers and water lilies, across the miles of cornfields and citrus orchards to my father's garden, flowering plants have come to rule the worlds of botany and agriculture. They also reign over an ethereal realm sought by artists, poets, and everyday people in search of inspiration, solace, or the simple pleasure of beholding a blossom.

"Before flowering plants appeared," says Dale Russell, a paleontologist with North Carolina State University and the State Museum of Natural Sciences, "the world was like a Japanese garden: peaceful, somber, green; inhabited by fish, turtles, and dragonflies. After flowering plants, the world became like an English garden, full of bright color and variety, visited by butterflies and honeybees. Flowers of all shapes and colors bloomed among the greenery."

THAT DRAMATIC CHANGE represents one of the great moments in the history of life on the planet. What allowed flowering plants to dominate the world's flora so quickly? What was their great innovation?

Botanists call flowering plants angiosperms, from the Greek words for "vessel" and "seed." Unlike conifers, which produce seeds in open cones, angiosperms enclose their seeds in fruit. Each fruit contains one or more carpels, hollow chambers that protect and nourish the seeds. Slice a tomato in half, for instance, and you'll find carpels. These structures are the defining trait of all angiosperms and one key to the success of this huge plant group, which numbers some 235,000 species.

Just when and how did the first flowering plants emerge? Charles Darwin pondered that question, and paleobotanists are still searching for an answer. Throughout the 1990s discoveries of fossilized flowers in Asia, Australia, Europe, and North America offered important clues. At the same time the field of genetics brought a whole new set of tools to the search. As a result, modern palebotany has undergone a boom not unlike the Cretaceous flower explosion itself.

Now old-style fossil hunters with shovels and microscopes compare notes with molecular biologists using genetic sequencing to trace modern plant families backward to their origins. These two groups of researchers don't always arrive at the same birthplace, but both camps agree on why the quest is important.

"If we have an accurate picture of the evolution of a flowering plant," says Walter Judd, "then we can know things about its structure and function that will help us answer certain questions: What sorts of species can it be crossed with? What sorts of pollinators are effective?" This, he says, takes us toward ever more sensible and productive methods of agriculture, as well as a clearer understanding of the larger process of evolution.

Elizabeth Zimmer, a molecular biologist with the Smithsonian Institution, has been rethinking that process in recent years. Zimmer has been working to decipher the genealogy of flowering plants by studying the DNA of today's species. Her work accelerated in the late 1990s during a federally funded study called Deep Green, developed to foster coordination among scientists studying plant evolution.

Zimmer and her colleagues began looking in their shared data for groups of plants with common inherited traits, hoping eventually to identify a common ancestor to all flowering plants. Results to date indicate that the oldest living lineage, reaching back at least 130 million years, is Amborellaceae, a family that includes just one known species, *Amborella trichopoda*. Often described as a "living fossil," this small woody plant grows only on New Caledonia, a South Pacific island famous among botanists for its primeval flora.

But we don't have an *Amborella* from 130 million years ago, so we can only wonder if it looked the same as today's variety. We do have fossils of other extinct flowering plants, the oldest buried in 130-million-year-old sediments. These fossils give us our only tangible hints of what early flowers looked like, suggesting they were tiny and unadorned, lacking showy petals. These no-frill flowers challenge most notions of what makes a flower a flower.

To see what the first primitive angiosperm might have looked like, I flew to England and there met paleobotanist Chris Hill, formerly with London's Natural History Museum. Hill drove me through rolling countryside to Smokejacks Brickworks, a quarry south of London. Smokejacks is a hundred-foot-deep hole in the ground, as wide as several football fields, that has been offering up a lot more than raw material for bricks. Its rust-colored clays have preserved thousands of fossils from about 130 million years ago. We marched to the bottom of the quarry, got down on our hands and knees, and began digging.

Soon Hill lifted a chunk of mudstone. He presented it to me and pointed to an imprint of a tiny stem that terminated in a rudimentary flower. The fossil resembled a single sprout plucked from a head of broccoli. The world's first flower? More like a prototype of a flower, said Hill, who made his initial fossil find here in the early 1990s. He officially named it *Bevhalstia pebja*, words cobbled from the names of his closest colleagues.

Through my magnifying glass the *Bevhalstia* fossil appeared small and straggly, an unremarkable weed I might see growing in the water near the edge of a pond, which is where Hill believes it grew.

"Here's why I think it could be a primitive flowering plant," said Hill. "*Bevhalstia* is unique and unassignable to any modern family of plants. So we start by comparing it to what we know." The stems of some modern aquatic plants share the same branching patterns as *Bevhalstia* and grow tiny flower buds at the ends of certain branches. *Bevhalstia* also bears a striking resemblance to a fossil reported in 1990 by American paleobotanists Leo Hickey and Dave Taylor. That specimen, a diminutive 120-million-year-old plant from Australia, grew leaves that are neither fernlike nor needlelike. Instead they are inlaid with veins like the leaves of modern flowering plants.

More important, Hickey and Taylor's specimen contains fossilized fruits that once enclosed seeds, something Hill hopes to find associated with *Bevhalstia*. Both plants lack defined flower petals. Both are more primitive than the magnolia, recently dethroned as the earliest flower, although still considered an ancient lineage. And both, along with a recent find from China known as *Archaefructus*, have buttressed the idea that the very first flowering plants were simple and inconspicuous.

LIKE ALL PIONEERS, early angiosperms got their start on the margins. In a world dominated by conifers and ferns, these botanical newcomers managed to get a toehold in areas of ecological disturbance, such as floodplains and volcanic regions, and adapted quickly to new environments. Fossil evidence leads some botanists to believe that the first flowering plants were herbaceous, meaning they grew no woody parts. (The latest genetic research, however, indicates that most ancient angiosperm lines included both herbaceous and woody plants.) Unlike trees, which require years to mature and bear seed, herbaceous angiosperms live, reproduce, and die in short life cycles. This enables them to seed new ground quickly and perhaps allowed them to evolve faster than their competitors, advantages that may have helped give rise to their diversity.

While this so-called herbaceous habit might have given them an edge over slow-growing woody plants, the angiosperms' trump card was the flower. In simple terms, a flower is the reproductive mechanism of an angiosperm. Most flowers have both male and female parts. Reproduction begins when a flower releases pollen, microscopic packets of genetic material, into the air. Eventually these grains come to rest on another flower's stigma, a tiny pollen receptor. In most cases the stigma sits atop a stalklike structure called a style that protrudes from the center of a flower. Softened by moisture, the pollen grain releases proteins that chemically discern whether the new plant is genetically compatible. If so, the pollen grain germinates and grows a tube down through the style and ovary and into the ovule, where fertilization occurs and a seed begins to grow.

Casting pollen to the wind is a hit-or-miss method of reproduction. Although wind pollination suffices for many plant species, direct delivery by insects is far more efficient. Insects doubtless began visiting and pollinating angiosperms as soon as the new plants appeared on Earth some 130 million years ago. But it would be another 30 or 40 million years before flowering plants grabbed the attention of insect pollinators by flaunting flashy petals.

"Petals didn't evolve until between 90 and 100 million years ago," said Else Marie Friis, head of paleobotany at the Swedish Natural History Museum on the outskirts of Stockholm. "Even then, they were very, very small."

A thoughtful woman with short brown hair and intense eyes, Friis oversees what many experts say is the most complete collection of angiosperm fossils gathered in one place. The fragile flowers escaped destruction, oddly enough, thanks to the intense heat of long-ago forest fires that baked them into charcoal.

Friis showed me an 80-million-year-old fossil flower no bigger than the period at the end of this sentence. Coated with pure gold for maximum resolution under an electron microscope, it seemed to me hardly a flower. "Many researchers had overlooked these tiny, simple flowers," she said, "because you cannot grasp their diversity without the microscope."

So we squinted through her powerful magnifier and took a figurative walk through a Cretaceous world of tiny and diverse angiosperms. Enlarged hundreds or thousands of times, Friis's fossilized flowers resemble wrinkled onion bulbs or radishes. Many have kept their tiny petals clamped shut, hiding the carpels within. Others reach wide open in full maturity. Dense bunches of pollen grains cling to each other in gnarled clumps.

Sometime between 70 and 100 million years ago the number of flowering plant species on Earth exploded, an event botanists refer to as the "great radiation." The spark that ignited that explosion, said Friis, was the petal.

"Petals created much more diversity. This is now a widely accepted notion,"

Friis said. In their new finery, once overlooked angiosperms became standouts in the landscape, luring insect pollinators as never before. Reproduction literally took off.

Interaction between insects and flowering plants shaped the development of both groups, a process called coevolution. In time flowers evolved arresting colors, alluring fragrances, and special petals that provide landing pads for their insect pollinators. Uppermost in the benefits package for insects is nectar, a nutritious fluid flowers provide as a type of trading commodity in exchange for pollen dispersal. The ancestors of bees, butterflies, and wasps grew dependent on nectar, and in so doing became agents of pollen transport, inadvertently carrying off grains hitched to tiny hairs on their bodies. These insects could pick up and deliver pollen with each visit to new flowers, raising the chances of fertilization.

INSECTS WEREN'T the only obliging species to help transport flowering plants to every corner of the Earth. Dinosaurs, the greatest movers and shakers the world has ever known, bulldozed through ancient forests, unwittingly clearing new ground for angiosperms. They also sowed seeds across the land by way of their digestive tracts.

By the time the first flowering plant appeared, plant-eating dinosaurs had been around for a million centuries, all the while living on a diet of ferns, conifers, and other primordial vegetation. Dinosaurs survived for another 65 million years, and some scientists think this was plenty of time for the big reptiles to adapt to a new diet that included angiosperms.

"Just before the dinosaurs disappeared, I think a lot of them were chowing down on flowering plants," says Kirk Johnson of the Denver Museum of Nature and Science. Johnson has unearthed many fossils between 60 and 70 million years old from sites across the Rocky Mountain region. From them he deduces that hadrosaurs, or duck-billed dinosaurs, subsisted on large angiosperm leaves that had evolved in a warm climatic shift just before the Cretaceous period ended. Referring to sediments that just predate the dinosaur extinction, he said, "I've only found a few hundred samples of nonflowering plants there, but I've recovered 35,000 specimens of angiosperms. There's no doubt the dinosaurs were eating these things."

Early angiosperms were low-growing, a fact that suited some dinosaurs better than others. "Brachiosaurs had long necks like giraffes, so they were poorly equipped for eating the new vegetation," says Richard Cifelli, a paleontologist with the University of Oklahoma. "On the other hand ceratopsians and duck-billed dinosaurs were real mowing machines." Behind those mowers angiosperms adapted to freshly cut ground and kept spreading.

Dinosaurs disappeared suddenly about 65 million years ago, and another group of animals took their place—the mammals, which greatly profited from the diversity of angiosperm fruits, including grains, nuts, and many vegetables. Flowering plants, in turn, reaped the benefits of seed dispersal by mammals.

"It was two kingdoms making a handshake," says David Dilcher, a paleobotanist with the Florida Museum of Natural History. "I'll feed you, and you take my genetic material some distance away."

Eventually humans evolved, and the two kingdoms made another handshake. Through agriculture angiosperms met our need for sustenance. We in turn have taken certain species like corn and rice and given them unprecedented success, cultivating them in vast fields, pollinating them deliberately, consuming them with gusto. Virtually every non-meat food we eat starts as a flowering plant, while the meats, milk, and eggs we consume come from livestock fattened on grains—flowering plants. Even the cotton we wear is an angiosperm.

Aesthetically, too, angiosperms sustain and enrich our lives. We've come to value them for their beauty alone, their scents, their companionship in a vase, a pot, on Valentine's Day. Some flowers speak an ancient language where words fall short. For these more dazzling players—the orchids, the roses, the lilies—the world grows smaller, crisscrossed every day by jet-setting flowers in the cargo holds of commercial transport planes.

"We try to deliver flowers anywhere in the world within 24 hours of when they're cut," said Jan Lanning, a senior consultant with the Dutch Floricultural Wholesale Board, the world's turnstile for ornamental flowers. "The business has really globalized."

On my way home from Friis's lab in Sweden, I had stopped in the Netherlands, the world's largest exporter of cut flowers. I asked Lanning to try to explain the meaning of his chosen work. He leaned forward with a ready answer.

"People have been fascinated by flowers as long as we've existed. It's an emotional product. People are attracted to living things. Smell, sight, beauty are all combined in a flower." He smiled at an arrangement of fragrant lilies on his desk. "Every Monday a florist delivers fresh flowers to this office. It is a necessary luxury."

Later that day in Amsterdam's Van Gogh Museum I spied a group of admirers crowded before a painting. I made my way there and pressed in among them. Suddenly I was staring at *Sunflowers,* one of van Gogh's most famous works. In the painting the flowers lean out of a vase, furry and disheveled. They transported me to my barefoot youth at the edge of my dad's garden on a humid summer evening alive with fireflies and the murmur of cicadas.

The crowd moved on, and I was alone with *Sunflowers*. My quest had come to this unexpected conclusion, an image of the first flower I can remember. Did van Gogh elevate the flower to an art form, or did the flower harness van Gogh's genius to immortalize itself in oils and brushstrokes? Flowering plants have conquered more than just the land. They have sent roots deep into our minds and hearts. We know we are passing through their world as through a museum, for they were here long before we arrived and may well remain long after we are gone.

SUSAN MILIUS

Why Turn Red?

FROM *SCIENCE NEWS*

Sometimes the simplest questions have the most interesting answers. Why do some leaves turn such bright, dazzling shades of red in the autumn? Susan Milius catches up with a young botanist in search of the answer—and whose modern-day research is reviving an idea first proposed in the nineteenth century.

A leaf turning red in the fall makes for a much greater mystery than a leaf turning yellow does. The yellowing signals simply a dropping of veils because the yellow pigment has lain hidden in the leaf during its long, green summer. When summer ends and the green pigments break down, the yellow shines through. Reds, however, don't loll around all summer. A leaf with only a few weeks left to hang on its tree summons its faltering resources for a burst of bright-red-pigment making.

What a time to redecorate. Cell physiologists have found a world inside an autumn leaf that resembles the pandemonium on a sinking ship. Metabolic pathways start to fail. Compounds break apart. Doomed cells rush to salvage the valuables, especially nitrogen, by sending them off to safer tissues. So in this final crisis, why make a special effort to turn red? Does the red-making machine turn on by accident, or do the red pigments contribute something valuable? Why would passengers fleeing the *Titanic* stop to repaint their staterooms?

There are plenty of proposed explanations, says David Lee, a tropical botanist at Florida International University in Miami. He and other pigment researchers say that modern analytical techniques are enabling them to test these ideas in new ways—and finally get some answers. The most abundant evidence, he says, has revived a 19th-century notion that the red pigments called anthocyanins serve as a protective device for faltering photosynthetic chemistry.

Red Start

IRONICALLY, Lee didn't get interested in anthocyanins until a job took him to a place without fall. He grew up with the humdrum autumn colors of the relatively dry landscape of eastern Washington State. "There wasn't much of an autumn show—a few trees in town turned red," he says. However, in 1973, he left temperate seasons behind when he joined the faculty at the University of Malaya in Kuala Lumpur.

Some of the tropical trees there burst into astonishing reds, though not all at the same time or for the same reason as each other. The Indian almond, for example, blushes brightly just before it sheds its leaves. The leaves of mangos and cacaos do the reverse, turning scarlet when they first sprout.

"A whole tree will quickly flush red," Lee says. "I saw it and thought, 'Wow, what's happening here?' "

Anthocyanins provide the red special effects for much of the plant kingdom. Their fireworks intrigued 19th-century biologists, who discussed the possibility that a leaf might make anthocyanins during a period of vulnerability, to shield the green chlorophyll pigments from sunburn. However, these intensely colorful compounds showed up in little walled-off pockets called vacuoles within cells. Since physiologists have often considered the vacuole "the cell's trash bag," says Lee, the sunscreen proposal faded into disfavor. For much of the past century, he says, physiologists classed anthocyanins as just some more trash.

Lee suspected the old idea might have something to it, perhaps in the screening of especially vulnerable leaves—the extremely young and the extremely old—from ultraviolet (UV) radiation. Yet anthocyanins have turned out not to absorb UV as well as some of their own chemical precursors in the leaf do. Making anthocyanins would actually deplete the store of better UV absorbers. "I became disenchanted with that hypothesis," Lee recalls, but he still wondered whether anthocyanins might shield a vulnerable leaf from some other menace.

In 1992 at a botanists' meeting in Hawaii, he met plant physiologist Kevin Gould of the University of Auckland in New Zealand. Over a breakfast in Woolworth's, they plotted a test of the sunscreen hypothesis using shade-loving species as examples of light-sensitive plants.

The two researchers focused on certain little plants that dot shaded forest floors and grow leaves with green tops and red undersides. For example, the common trout lily of northeastern forests does this, as do some begonias.

Lee had found a Malaysian begonia and a Costa Rican melastome that naturally vary in leaf color, some individuals sprouting all-green leaves and others putting out leaves with red undersides. Lee and Gould blasted samples of all these leaves with intense light. Physiologists had already shown that such blasts overload light-gathering chlorophyll and slow it down, a misfortune called photoinhibition.

In Lee and Gould's experiment, all-green leaves seemed to suffer greater photoinhibition than did two-tone ones of the same species. Reporting their finding in 1995, the two physiologists proposed that random strikes of bright sunlight on the light-dappled forest floor could pose great dangers. A plant with a little protection in the form of anthocyanins could off-load some of that sudden excess energy in the form of its chlorophyll and better withstand a blast.

Red Spread

IN THE 1990S, other researchers also explored the idea of red pigments as sunscreen. Debate bloomed over how to devise a test that avoids confounding factors, such as different rates of photosynthesis in different-colored leaves.

In 1999, researchers at the University of Queensland in Australia refined the bright-light tests performed by scientists including Lee and Gould. In an experiment on the tropical *Bauhinia variegata,* Robert C. Smillie and Suzan E. Hetherington flashed an assortment of its red or green pods with bursts of white, blue-green, or red light. The red pods tolerated the white and blue-green light flashes better than the green pods did. Yet the red pods didn't show any superior tolerance to bursts of red light. The researchers contended that in the latter case, anthocyanins, which can't soak up red wavelengths, weren't protecting the chlorophyll.

Lee then joined Taylor S. Feild and N. Michele Holbrook of Harvard University in a similar experiment on autumn leaves. The researchers chose red-osier dogwood shrubs because they end the year in multiple colors. In fall, leaves bathed in brilliant sunlight turn red, but shaded leaves don't develop anthocyanins and so just turn yellow. The red leaves recovered faster from flashes of intense blue light, the researchers reported in the October 2002 *Plant Physiology*. Flashes of red light, the wavelength that anthocyanins can't absorb, had about the same effect on red leaves as on yellow ones.

The finding dovetails with physiological studies from other labs that suggest that leaves may need special protection during their final weeks. Tests showed that old leaves are more vulnerable to photoinhibition than younger but mature ones are. In a color-changing leaf, the plant's metabolic pathways for making the initial capture of energy don't lose their efficiency as fast as the subsequent pathways for processing that energy do, a risky imbalance that invites overloads. Seasonal stresses, such as chilling temperatures, also hobble the leaf metabolism.

Yet during autumn, the aging leaf has to salvage as much nitrogen as possible and send it to tissues that will survive the winter. So, as decrepit as the photosynthetic mechanism becomes at the end, it has to keep catching and processing sunlight if the leaf is to finish the salvage operation.

That scenario prompted William A. Hoch of the University of Wisconsin–Madison to look at the geographic history of intense red color. He hypothesized that plants would be most likely to manufacture anthocyanins in climes where temperatures often plunge during autumn. So, he ranked the intensity of anthocyanins in fall-coloring in nine genera of woody plants. Some of these were native to either a cold zone in Canada and the northern United States, others to a milder, maritime clime in Europe. Out of 74 species, the 41 that flamed out with reddest leaves all came from the North American chilly zone, he reported in the January 2001 *Tree Physiology*.

Blueberries, Etc.

THE EVIDENCE HAS BEEN building nicely for anthocyanins as safety measures against light overdose, according to Gould. Yet he doesn't expect that to be their only function. "They're very talented molecules," he says.

He got the urge to test for another benefit, he says, while reading a newspaper article touting the health benefits of diets that include blueberries. Antioxidant pigments abound in blueberries, and Gould decided to explore whether the antioxidizing powers of the leaf anthocyanins that he was studying benefit their plants.

When purified in the lab, these pigments sop up free radicals, which are alarmingly energized substances that can damage DNA, proteins, and membranes. Anthocyanins in a test tube can corral free radicals four times as well as do the well-known antioxidants vitamin C and E, says Gould. He started planning a test for anthocyanins' antioxidant effects inside a living plant.

"It took us a long time," he says, "but I had some very diligent students." They borrowed an imaging technique called epifluorescence microscopy from research on animal cells. With it, they could watch bursts of the oxidizing agent hydrogen peroxide as it was released in a cell. To observe the actions of antioxidants, the researchers had to figure out a way to trigger such oxidizing bursts in plant cells.

Gould remembered that one of their study subjects, a New Zealand piebald shrub called *Pseudowintera colorata,* developed small red pimples on its leaves where aphids pricked them to suck sap. When the researchers stabbed the leaves with a very fine needle, they triggered bursts of hydrogen peroxide in cells. A steady-handed scientist could induce the bursts and the subsequent redness as well as an aphid does. "We could write the word 'red,' and it came out red two days later," says Gould.

After patiently perfecting these techniques, Gould's lab made a movie. The researchers filmed the stabbing of both the all-green and the red-splotched leaves of *P. colorata.* In the October 2002 *Plant, Cell and Environment,* Gould and his colleagues reported seeing an oxidative burst of hydrogen peroxide a minute or less after they pushed the needle into the upper layers of leaf tissue. In red tissues, the burst faded quickly. In green ones, however, it intensified, and hydrogen peroxide concentrations soared for at least 10 minutes. Gould contends that anthocyanins are the compounds most likely to have quenched the oxidative burst.

Lee welcomes the report enthusiastically. "It's the first evidence [for antioxidant behavior] in a living plant," he says.

A suggestion for yet a third function for anthocyanins in leaves comes from physiologist Linda Chalker-Scott of the University of Washington in Seattle. She proposes that the pigments regulate water movement. She has contributed a chapter on the idea to the book *Anthocyanins in Leaves* (Kevin Gould and David Lee, eds., Academic Press 2002).

Anthocyanins dissolve in water, whereas chlorophyll and a lot of other cell pigments don't, she explains. Water loaded with any dissolved substance has what physiologists call lower osmotic potential, a decreased tendency to flow away. Loading water with dissolved substances also lowers the temperature at which water freezes, potentially an advantage on a frosty fall night.

Chalker-Scott points out that many plants blush red at water-related stresses such as drought, salt buildup, and heat. Her experiments testing the idea have been largely on hold since 2001, when the building housing her lab was firebombed during a protest targeting another researcher's genetic engineering project.

Plenty of other ideas for anthocyanins' function also remain to be tested. Observers of fungus-farming ants, for example, reported in the 1970s that the ants avoid taking red leaves home to feed to their garden. Researchers have speculated that anthocyanins might discourage growth of some fungi.

Another hypothesis states that anthocyanins keep leaves from overheating; an alternative has the pigments protecting leaves from cold. Gould notes that a birch species he encountered in Finland holds on to its red leaves year-round, despite temperatures that plunge to −40°C.

In 2001, a paper by the late theorist W. D. Hamilton and Samuel P. Brown of the University of Montpellier in France mused about whether autumn coloring shares a communication role with the peacock's tail. The healthiest birds can grow the most spectacular tails, so a cruising female can get an accurate assessment of a prospective mate's health by checking out his plumage. In a similar way, Brown and Hamilton speculated that the healthiest trees might put on the flashiest fall displays. This leaf signal might give fall-active predatory insects, such as aphids, accurate information about which trees have good defenses and which ones might be easy pickings.

Even if none of these or the abundant other suggestions pans out, researchers already know enough to raise anthocyanins from the category of cellular trash to their deserved status as vital molecules. A big question still remains: If these pigments are so great, why don't all leaves turn red?

Thomas Eisner

The Mosquito's Buzz

FROM *WINGS*

In one of those odd bits of happenstance that frequently occur in the history of science, a man whose interests lay in deadly weapons may have discovered the delicate intricacies of insect acoustics. The biologist Thomas Eisner explains.

Hiram Maxim is best known as the inventor of a machine gun, the Maxim gun. He is given little credit for another, much more benign contribution to humanity, a discovery which was by nature serendipitous, but certainly more worthy of recognition than the lethal contraption that earned him a knighthood from Queen Victoria. Hiram Maxim, I would argue, was the founder (or at the least, co-founder) of insect bio-acoustics, a discipline now thriving, but hardly emergent in the late 1800s when Maxim made his discovery.

Maxim was an electrical engineer whose reputation was at a par with that of his contemporary, Thomas Alva Edison. He was in high demand for his talents, at the very time when electrification was coming into vogue. In 1878 Maxim was asked to install electricity in the Grand Union Hotel in Saratoga Springs, New York. This was an offer he could not possibly refuse, as Saratoga Springs was one of the primary resorts of the period, a spa where the munificent and famous could mingle, while taking in the waters and exchanging fortunes at the track. Frequenters of the spa over the years included Mark Twain,

Lily Langtry, Victor Herbert, "Diamond Jim" Brady, Lillian Russell, and count-less other celebrities. The Grand Union was the largest and best known of Saratoga's hotels. Its dining hall alone could seat over two thousand. And its nightly balls, in the great shaded outdoor gardens, were world-renowned. It was these gardens that Maxim was asked to illuminate, and he complied by in-stalling arc lights. He put the generators in place, wired things up in appropri-ate fashion, and on the designated night, turned on the switches. One can only gather, judging from the illustrated news accounts of the time depicting the richly attired clustered in awe around the shining fixtures, that the advent of the electric era in Saratoga Springs was received with acclamation.

Success did not keep Maxim from remaining observant. He was struck by a peculiar phenomenon. One of the generators, which was emitting a high-pitched sound when it was turned on, was attracting droves of insects during the evening hours. Mystified, Maxim availed himself of a loupe, and noted that the insects were all of one kind. With expert help he was able to ascertain that they were all male mosquitoes, and he ventured the guess that these had been attracted because the sound emitted by the generator was imitative of the buzz ordinarily given off by the female mosquito in flight. The buzz of the female, he reasoned, was the mosquito's mating call. The idea was new, he thought, and worth publishing. Scientist that he was, he put pen to paper, and proposed the notion in a manuscript that he submitted to a technical journal. At the time, the scientific establishment was resistive to innovative thought, and his paper was rejected. He eventually wrote up his observations in a letter that he sub-mitted to *The Times* of London, which published it on October 28, 1901. The date is a landmark of sorts in entomological history.

We know now that Maxim was right. Male mosquitoes as a rule are indeed attracted to the flight sounds of the female, and they have special ears, in the form of their antennae, for detection of the buzz. A simple experiment can be carried out by anyone with access to a cage of mosquitoes. Take a tuning fork of appropriate pitch (humming frequencies of three hundred to eight hundred cycles per second will do nicely), tap it so it will hum, and introduce it into the cage. You will note that the male mosquitoes will take to flight and aggregate around the fork. They are irresistibly attracted to the sound. While the fork is humming, you can draw the males from the cage and walk about with them, leading them by the fork until you are ready to return them into the cage. You will not lose males as long as the fork is vibrating. Love, one is tempted to muse, even in the world of mosquitoes, takes priority over freedom.

A more sophisticated experiment can be done by using live females in lieu of the fork. Female mosquitoes can be glued by the thorax to the end of a fine wire (using a small dab of wax), and they will "fly" when thus tethered, beating their wings in normal fashion. Introduce such a buzzing female into a cage with males, and she will attract them. Set up the female in front of a camera, and you may have the privilege of recording her amorous antics. Males will converge, singly or in groups, and eventually one of them will succeed in positioning himself belly-to-belly beneath the female, like a torpedo under a plane. The female may help the male secure his hold. Copulation sometimes follows without the female ceasing to beat her wings. Belly-to-belly is not the usual way for insects to mate—males tend to mount females in the insect world. In mosquitoes, the strategy may have evolved specifically to permit aerial coupling. Copulation runs its course in a matter of seconds in mosquitoes, attesting to the extraordinary speed with which sperm transfer can take place in insects. It takes longer in many insects, but tends to be kept short in species that incur risks when paired. In mosquitoes, which may copulate in midair, the strategy appears to be intended to "get things over with quickly."

Mosquitoes, on emergence from the pupa, are not instantly ready for "action." The males have a rear end that must first rotate 180 degrees before it lends itself to belly-to-belly coupling. They are "born" with the genital apparatus oriented the ancestral way, and need about forty-eight hours to twist the apparatus half a full circle. During this period the males are kept relatively insensitive to sound, and as a result resistant to the hum of temptresses nearby. Acoustical sensitivity in male mosquitoes varies in relation to the degree of deflection of the bristles on their antennae. The antennae have a swollen segment near the base, packed with sensory neurons that respond to the vibration of the antennal shaft. The antennae are especially prone to resonate in response to the sound frequencies emitted by the buzzing female. The bristles on the antennal shaft act as an amplifier system. When the bristles are erect, they help "collect" incoming sound, and the male is acoustically more sensitive. When they are recumbent against the shaft, the male is hearing-impaired. At emergence from the pupa the bristles of the male are in the recumbent state, hence the male's relative deafness. By the time the rear has undergone its twist, the bristles have become erect.

The male's antennae, with their bushy covering, can impose a drag in flight, which may be energetically costly. Remarkably, in some mosquitoes, the antennal bristles are kept in the recumbent state much of the day, and are erected only for a few hours at dusk, when the sexes are at play. The bristles in these mosquitoes are controlled by a circadian clock, which sets the rhythm of their erection. The clock resides in the thorax, and it is from the thorax that the neural signals arise that regulate the angular orientation of the bristles. The mechanism of bristle erection is itself interesting, in that it involves the controlled application of hydrostatic pressure at the hinged bases of the bristles.

Maxim might have enjoyed knowing that not all mosquitoes court on the wing. In *Opifex fuscus,* a New Zealand species, mating may begin before the female has taken to the air. The male *Opifex* routinely patrols the air space above waters likely to contain pupae. He has good eyesight and is apparently able to detect ripples created as pupae surface for air. When he spots a pupa, he seizes it and remains in attendance as the emergence takes place. He may physically assist in the extrication process, and if the emergent mosquito is a female, may attempt to clasp her genitalia before she is even out and about.

Mosquitoes are not everyone's favorite insects, although they are of immense interest. As pests and vectors of disease they have rightfully commanded considerable attention, but there are doubtless many species, including undiscovered ones, from which we have an enormous amount to learn. It is also worth considering the many direct benefits that mosquitoes bring to the environments and ecosystems in which they live. As larvae they are an important food source for fish, and as adults they are similarly useful to other insects and birds. The adult males are nectar feeders and significant pollinators in some arctic plant communities.

While mosquitoes live on, the Grand Union Hotel does not. It was razed in 1952, to make way for a supermarket. Grand Union by name, the market was eventually to proliferate into a chain. The hotel has long since been forgotten. Hiram Maxim is remembered mainly for his gun.

LAWRENCE OSBORNE

Got Silk

FROM *THE NEW YORK TIMES MAGAZINE*

A genetically modified goat that can produce the silk of spiders in its milk may seem like the stuff of science fiction, but thanks to the biotechnology revolution, it's a reality. The journalist Lawrence Osborne tours a farm that may offer a preview of our transgenic future.

As soon as I walk into the humid goat shed in my Tyvek suit and sterilized boots, a dozen Nubians run up to the fence and begin sniffing at me, their Roman noses dilated with fervent curiosity. "They're a little frisky," a technician explains, shooing them back toward their playpen toys. "It's artificial insemination time, you know."

The technician, a young woman in galoshes named Annie Bellemare, and two colleagues are playing a trick on a long-bearded billy goat. Leading him up to a female in heat, they let him mount her; but at the last moment, they whip out a warmed, rubber-lined bottle and have him discharge into it. "There," they cry, holding up a phial of goat semen. "Good boy!"

I look around the pen. Hundreds of sly-looking, inquisitive goats are staring at me intently. They seem unexceptional enough, but the goats that are being bred here are far from ordinary. This is a so-called transgenic farm—a place where animal species are either cloned or genetically mixed to create medically useful substances—owned and run by a firm named Nexia Biotechnologies. It is housed on a former maple-sugar farm in rural Quebec, not far from the remote hamlet of St.-Telesphore. Nexia's facility is one of only three transgenic farms in the world. (One of the company's rivals, PPL Therapeutics, runs the farm in Scotland that collaborated in the production of the famous sheep clone, Dolly.)

Out here in this tough French-speaking farming country, however, hardly anybody gets worked up about the fact that on the old St.-Telesphore sugar farm, a new chapter in biotechnology is being written. Nexia scientists are pursuing a bizarre experiment straight out of *The Island of Dr. Moreau*, H. G. Wells's dark science-fiction fable of a mad scientist who breeds experimental animals on his private preserve.

"Oh, it's not that weird," Nexia's president and CEO, Jeffrey D. Turner, says as we walk around the pens, being nibbled constantly by aroused goats. "What we're doing here is ingeniously simple," he says. "We take a single gene from a golden orb-weaving spider and put it into a goat egg. The idea is to make the goat secrete spider silk into its milk."

Milk silk?

Turner, a bouncy 43-year-old scientist turned biotech entrepreneur, makes a sweeping gesture at his bleating production units. "Spider silk is practically the world's strongest material," he explains. "It's much stronger than steel—five times as strong. We're going to make fishing lines out of it."

I raise my eyebrows dubiously.

"Yes. Biodegradable fishing lines. Or maybe tennis racket strings." He grows even more animated. "You could make hundreds of things out of spider silk, if only you could produce enough of it. Biodegradable sutures for surgery . . . replacement ligaments or tendons . . . hemostatic dressings . . . fashion. We call our product BioSteel."

Turner isn't simply fantasizing. Nexia foresees tapping into the $500 million markert for fishing materials as well as the $1.6 billion market for industrial fibers in the near future. And the haute-couture world is already intrigued by a nearly weightless gossamer-like fabric. But the real gold mine might be body armor: the Pentagon is working with Nexia to develop a prototype of a new kind of vest that might be made entirely out of goat silk. The vest would be only a little thicker than nylon, but it could stop a bullet dead.

"It's nothing short of a revolution," Turner exclaims. "This special silk is the first transgenic material ever made. The amazing thing, however, is that we're changing the world from a tiny low-rent sugar farm, and our only machinery is a goat."

Turner is very affectionate with his goats. A number of different species are being tested for the spider-gene project. In one pen a gang of floppy-eared Nubians frolic and duel, raising themselves on their back legs and then clashing foreheads. Next door live the Saanens from Switzerland, all of them white, rather meeker and well mannered, quietly cocking their heads at the sound of human voices. Across the way stand a dozen West African Dwarf goats (once used by the Hamburg Zoo as food for big cats from Africa).

"We use West African Dwarf goats because they're sexually active all year round," Turner says. "Unlike American goats, which are only active in the fall and spring." He winks. "The African goats get sexually mature in three months. This helps reach the output potential quicker."

Turner once again admires his flock. "You could call them Spidergoats," he says. "But that would give people misconceptions. They're only 1/70,000th spider, after all. When it comes down to it, they're just normal goats with one spider gene in them. They're just goats." He pauses. "Mostly."

SCIENTISTS HAVE BEEN tinkering with the DNA of animals for years. Researchers have inserted into rhesus monkeys the gene that makes jellyfish glow in the dark; they've produced chickens that never grow feathers. But only recently have they begun to develop large-scale industrial plans for these creatures. For example, a company in Georgia called ProLinia has cloned cattle and hogs to produce more genetically desirable breeding stock. After scientists at Johns Hopkins produced enormous "supermice" by removing the gene that limits muscle growth, researchers have scrambled to create the same results in sheep, pigs and chickens.

Inevitably, some bioethicists are alarmed by these projects. And Turner agrees that some of these experiments are creepy.

"Why do we need cloned sheep?" he asks. "What the hell's the use of millions of cloned sheep? Dolly was a scientific stunt." He tells me that Nexia's project is less about altering nature than harnessing it. The company's goal isn't to create weird goats; they're merely a means of producing useful quantities of spider silk, a simple substance created eons ago by natural evolution. Turner says that what Nexia is really up to isn't mere genetic engineering, it's "biomimicry."

In her 1997 book, *Biomimicry*, Janine M. Benyus observed that while humans create synthetic materials by means of high temperatures and pressures ("heat, beat and treat" methods, as they are known), nature does so under life-friendly conditions. That is to say, in water, at room temperature and without harsh chemicals. "Nature's crystals are finer, more densely packed, more intricately structured and better suited to their tasks than our ceramics and metals are suited to ours," Benyus observes. Inspired by this, materials scientists are now looking to merge biology and engineering—the natural and human-made.

"In the future, animals will be our factories," Turner says as we plod through the facility. "Very cheap factories."

This is a land of silos and bleached cherry-red barns, somewhere between the St. Lawrence and Ottawa Rivers. "We need to be where people aren't," Turner explains. Nexia's converted *cabane a sucre* and the surrounding land, purchased five years ago from a local farmer, look sweetly ordinary. But the new facilities are meticulously decontaminated. The company's corporate headquarters are just 15 miles down the road, rising from the flatlands of Vaudreuil-Dorion like a futuristic castle keep. Inside, the corridors are freshly carpeted and sunlit; the labs are shiny and uncluttered and stocked with the latest gadgets. These labs are known as "Class-100,000 rooms," which means that each cubic yard of air contains less than 100,000 motes of dust. Staff members proudly show me the latest PCR (polymerase chain reaction) machines—the photocopiers of the gene world—that look like high-tech adding machines. Pinned to the walls are some curious images derived from what is known as FISH analysis. (The acronym stands for Fluorescent In Situ Hybridization.) These images show the goat genes as ghostly strands of dark orange, inside which one can clearly see the bright yellow segments of alien spider silk genes. Nearby are cute pinups of Nexia's original four transgenic goats, Willow, Bay, Santiago and Zeus.

Nexia used cloning to make its four founder animals, though the descendant animals are allowed to breed sexually. One pic shows Willow, Canada's first transgenic farm animal, posing coquettishly on a little orange plastic bobbin. I am told that she is, in fact, 1/70,000th human. This is because she has been specifically engineered to manufacture proteins for use in medical drugs like clot-busters, another source of income for Nexia. I look at her closely. Am I going mad or do I detect a human gleam in her eye?

HOW DOES a spider gene get into goat milk in the first place? Nexia uses two common spider specimens, *Araneus diadematus* (the common garden spider) and *Nephila clavipes* (the golden orb weaver, native to many tropical forests). The spiders are frozen in liquid nitrogen, then ground into a brown powder. Since every cell of a spider contains the precious silk-producing genes, it's easy to extract them. These genes are then tested in the "Charlotte machine," what Turner calls a "synthetic goat" that tests whether or not the gene will function inside an actual goat.

Next, the gene is altered. A "genetic switch" is added, which programs the gene to "turn on" only inside the mammary gland of its new female host during lactation. The altered gene is then pushed on a fine glass pipette into a goat egg. The baby goat will have a spider gene present in each of its cells (its eyes, ears and hooves will all be part spider), but only in the mammary glands of female goats will the silk gene actually spring to life. The goat will eventually start lactating a kind of silk-milk mixture, which looks and tastes just like normal milk.

This milk is first skimmed of fat, and salt is added to make the silk proteins curdle into thin whitish particles that promptly sink to the bottom. After the residue has been removed from the milk, a little water is added to this sediment until it turns into a golden-tinged syrup. This silk concentrate is known to scientists as "spin dope" and is more or less identical to what is inside a spider's belly. Now completely stripped from its milky context, the syrupy raw silk is ready for spinning.

Nexia's labs are packed with odd machines that replicate a spider's anatomy. First there is an extrusion machine, a strange-looking three-foot-tall apparatus bristling with aluminum pipes, designed to force the raw silk material through a tiny hole. As the silk comes out through this aperture, it is immediately stretched inside a long steel bathtub—at full tilt, roughly a hundred yards of it an hour.

Then the silk, which is transparently shiny with a white tinge, is taken to a spinner and strung out between two spindles a yard apart, which stretch the threads out as finely as possible. The idea is to do what a spider does naturally: subject the silk to tremendous stretching, or "shearing." This not only elongates it but actually strengthens the material as well. After being spun and wound around a plastic bobbin, some of the threads are then passed to a tensile tester, which measures their strength. In the production room, Turner hands me a few 20-micron-wide strands, frail as gossamer. The difficulty, he says, is making the silk as evenly as a spider does.

As we pass through yet more rooms filled with liquid nitrogen tanks where frozen goat semen and ova are stored, Turner explains to me the enigmatic inner world of spiders and their miraculous silk and their connection to modern needs.

Four hundred million years ago, he begins excitedly, spiders were doing just fine as ground hunters until one day bugs started flying. "The spider's evolution comes out of a kind of arms race between spiders and bugs. The bugs start flying to get away from the spiders, so the spiders have to come up with a new weapon." Most spider species died out, but a few developed a new talent, namely, spinning webs. The silk had to both be invisible to a bug's vision and virtually indestructible. Only spiders capable of making superfine, powerful silk survived—a perfect example of evolutionary pressure.

What's special about spider silk, as opposed to silk from worms, is that it is a unique liquid crystal. And that's what's magical, says Turner. "Liquid crystals are the Holy Grail of material sciences. They make for incredibly tough, light, strong materials with phenomenal properties. It's way beyond anything we humans can make. Milled steel pales next to it."

But the complexity of arachnid silk is also what is problematic about it, from the point of view of biomimicry. Spider-silk proteins consist of very long strings of amino acids that are difficult to decode, and little is known of how spiders actually unravel them and spin them into threads. A spider, moreover, constructs its web methodically out of different kinds of silk. It builds diagonal support lines called "dragline silk" (which it also uses to hoist itself around its web) and then inner wheels called "the capture spiral" made from a more viscid, sticky silk. Dragline silk, says Turner, is the "best stopping material you've ever seen," but how it's actually made inside a small orb weaver's abdomen remains mysterious. And whereas spiders produce up to seven kinds of silk proteins, BioSteel, as yet, contains only one.

As a result, BioSteel doesn't have all the resistant strength of spider silk—yet. Part of the mystery of spider silk's tremendous strength, current research suggests, lies in the spinning rather than in the internal chemistry of the silk itself. It seems that the silk proteins self-assemble as they are squeezed out of the spider's glands much like toothpaste being squeezed out of a tube. The stretching spontaneously causes the proteins to line up and lock into each other. "That's why we've spent so much money on these extrusion machines," Turner says. "The secret is in the spinning."

In any case, the properties of spider silk have long been recognized. Fishermen in India have always prized it for the making of their nets; American Civil War soldiers frequently used it as a surgical dressing. The problem lay always in getting sufficient quantities of it. Whereas silkworms are peaceful herbivores and can easily be farmed, spiders are aggressive territorial carnivores that need plenty of space and solitude. In farm conditions, they moodily attack and eat each other.

Farming zillions of spiders, then, is far too tricky. But farming peaceable goats is a cinch. Yet how to get the desirable material from a rather nasty predator like a spider into the reproductive system of a kindly animal like a Nubian goat? Enter the odd subject of mammary glands.

The mammary gland is a perfect natural factory for the synthesizing and production of proteins. It occurred to Turner, who had been working on lactation at McGill University's animal sciences department in the mid-to-late '80s, that, theoretically, one could introduce foreign genes into an animal's mammary gland and get any given protein out of the animal without killing it, much as one milks a cow. Given the enormous expense of manufacturing drugs artificially, transgenic animals offered a brilliant way to make dirt-cheap drugs; $50,000 worth of proteins could be extracted from a few buckets of milk at a cost of about $12 of hay! The logic seemed irresistible: the udder as factory outlet.

In 1993, Turner was approached by the two venture-capitalist godfathers of Canada's budding biotech industry, Bernard Coupal and Ed Rygiel. They had heard of his work at McGill and were interested in finding a way to create a transgenic goat. But where most transgenics is concentrated on making drugs, Turner, Coupal and Rygiel eventually wondered if it might not be more practical, and less risky, to concentrate on materials. For one thing, they realized, it's almost impossible for small companies to manufacture drugs. But a simple material that doesn't need FDA approval is quite another thing. And when they considered the possible uses of spider silk, they were astounded.

"Humans never think about size," Turner says. "If an animal doesn't make stuff on a scale we understand, we just ignore it. But insects and marine animals, although they're tiny, make incredible materials that we could use. Who's to say we can't?"

NEXIA DOESN'T ONLY farm goats in St.-Telesphore. It also has ambitious plans to turn an old Air Force base on the American side of the border into its mass-production facility for BioSteel. As I approach this decommissioned base just outside Plattsburgh, New York, I look through the miles of lonely fencing at the old concrete bunkers where nuclear missiles were most likely housed. They rise from the ground like ancient tombs covered with grass. A few floppy-eared Nubian goats stand incongruously on top of them, wagging their tails and bleating.

Nexia's sympathetic farm manager, Thomas Ballma, tells me that the goats just love rolling down the grassy sides in summer. "We can't hardly control them," he says as he shows me the inside of a newly refurbished bunker coated with epoxy paint. Inside the 80-foot-long cave our voices echo ominously as he points out with some pride the new ventilation ducts and electric cables. Nexia is trying to breed as many goats here as it can. From the present 302 goats they hope to have 1,500 a year from now.

We wander into one of the inhabited bunkers, where dozens of mop-haired Angoras jump to attention. Then they come trundling over to us en masse, licking our hands and cocking their heads inquisitively. I remark that the country music playing on the loudspeakers is rather loud. Is that Dolly Parton?

"Oh, they love Dolly Parton," Ballma says. "Country music has the steadiest beat. It keeps them calm and happy. Heavy metal, though, gets them agitated."

A shipment of goats has just arrived from Georgia, and as we stroll around the gigantic half-abandoned base, Ballma tells me how Nexia has revitalized the sagging post–Cold War economy of Plattsburgh. "It's been a godsend," he admits. "Even though it seems a little improbable. I've been raising goats for years, I love them, so at first the idea of making them secrete spider silk kind of weirded me out. But now I understand it. It's not what people think."

"Not Dr. Moreau?" I ask.

"No! We're just making fishing nets here. It's pretty normal, really."

As we stand in the old air-control tower overlooking the base I can hear a faint bleating of happy goats. From nuclear bombs to transgenic goats, it seems a strange progression, I say.

"Sure," he replies. "But perhaps it's just our own cleverness that weirds us out."

BRENDAN I. KOERNER

Disorders Made to Order

FROM *MOTHER JONES*

Feel shy? Tense? Frightened? There's a pill for you. As large pharmaceutical companies compete to cash in on the next blockbuster like Prozac, they can make relatively rare disorders seem like major epidemics. Tracking the full-scale media blitz that marked the introduction of one company's antianxiety medication, Brendan I. Koerner exposes the shaky science behind the marketing.

Word of the hidden epidemic began spreading in the spring of 2001. Local newscasts around the country reported that as many as 10 million Americans suffered from an unrecognized disease. Viewers were urged to watch for the symptoms: restlessness, fatigue, irritability, muscle tension, nausea, diarrhea, and sweating, among others. Many of the segments featured sound bites from Sonja Burkett, a patient who'd finally received treatment after two years trapped at home by the illness, and from Dr. Jack Gorman, an esteemed psychiatrist at Columbia University. Their testimonials were intercut with peaceful images of a woman playing with a bird, and another woman taking pills.

The disease was generalized anxiety disorder (GAD), a condition that, according to the reports, left sufferers paralyzed with irrational fears. Mental-health advocates called it "the forgotten illness." Print periodicals were awash in stories of young women plagued by worries over money and men. "Everything took 10 times more effort for me than it did for anyone else," one woman told the *Chicago Tribune*. "The thing about GAD is that worry can be a full-time job. So if you add that up with what I was doing, which was being a full-time achiever, I was exhausted, constantly exhausted."

The timing of the media frenzy was no accident. On April 16, 2001, the U.S. Food and Drug Administration (FDA) had approved the antidepressant Paxil, made by British pharmaceutical giant GlaxoSmithKline, for the treatment of generalized anxiety disorder. But GAD was a little-known ailment; according to a 1989 study, as few as 1.2 percent of the population merited the diagnosis in any given year. If GlaxoSmithKline hoped to capitalize on Paxil's new indication, it would have to raise GAD's profile.

That meant revving up the company's public-relations machinery. The widely featured quotes from Sonja Burkett, and the images of birds and pills, were part of a "video news release" the drugmaker had distributed to TV stations around the country; the footage also included the comments of Dr. Gorman, who has frequently served as a paid consultant to GlaxoSmithKline. On April 16—the date of Paxil's approval—a patient group called Freedom From Fear released a telephone survey according to which "people with GAD spend nearly 40 hours per week, or a 'full-time job,' worrying." The survey mentioned neither GlaxoSmithKline nor Paxil, but the press contact listed was an account executive at Cohn & Wolfe, the drugmaker's PR firm.

GlaxoSmithKline's modus operandi—marketing a disease rather than selling a drug—is typical of the post-Prozac era. "The strategy [companies] use—it's almost mechanized by now," says Dr. Loren Mosher, a San Diego psychiatrist and former official at the National Institute of Mental Health. Typically, a corporate-sponsored "disease awareness" campaign focuses on a mild psychatric condition with a large pool of potential sufferers. Companies fund studies that prove the drug's efficacy in treating the affliction, a necessary step in obtaining FDA approval for a new use, or "indication." Prominent doctors are enlisted to publicly affirm the malady's ubiquity. Public-relations firms launch campaigns to promote the new disease, using dramatic statistics from corporate-sponsored studies. Finally, patient groups are recruited to serve as the "public face" for the condition, supplying quotes and compelling human stories for the media; many of the groups are heavily subsidized by drugmakers, and some operate directly out of the offices of drug companies' PR firms.

The strategy has enabled the pharmaceutical industry to squeeze millions in additional revenue from the blockbuster drugs known as selective serotonin reuptake inhibitors (SSRIs), a family of pharmaceuticals that includes Paxil, Prozac, Zoloft, Celexa, and Luvox. Originally approved solely as antidepressants, the SSRIs are now prescribed for a wide array of heretofore obscure afflictions—GAD, social anxiety disorder, premenstrual dysphoric disorder. The proliferation of diagnoses has contributed to a dramatic rise in antidepressant sales, which increased eightfold between 1990 and 2000. Prozac alone has been used by more than 22 million Americans since it first came to market in 1988.

For pharmaceutical companies, marketing existing drugs for new uses makes perfect sense: A new indication can be obtained in less than 18 months, compared to the eight years it takes to bring a drug from the lab to the pharmacy. Managed-care companies also have been encouraging the use of medication, rather than more costly psychotherapy, to treat problems like anxiety and depression.

But while most health experts agree that SSRIs have revolutionized the treatment of mental illness, a growing number of critics are disturbed by the degree to which corporate-sponsored campaigns have come to define what qualifies as a mental disorder and who needs to be medicated. "You often hear: 'There are 10 million Americans with this, 3 million Americans with that,'" says Barbara Mintzes, an epidemiologist at the University of British Columbia's Centre for Health Services and Policy Research. "If you start adding up all those millions, eventually you'll be hard put to find some Americans who don't have such diagnoses."

WHEN PAXIL HIT the market in 1993, the drug's manufacturer, then known as SmithKline Beecham, lagged far behind its competitors. Eli Lilly's Prozac, the first FDA-approved SSRI, had already been around for five years, and Pfizer had beaten SmithKline to the punch with Zoloft's debut in 1992. With only a finite number of depression patients to target, Paxil's sales prospects seemed limited. But SmithKline found a way to set its drug apart from the other SSRIs: It positioned Paxil as an anti-anxiety drug—a latter-day Valium—rather than as a depression treatment.

SmithKline was especially interested in a series of minor entries in the Diagnostic and Statistical Manual of Mental Disorders (DSM), the psychiatric bible. Published by the American Psychiatric Association since the 1950s, the DSM is designed to give doctors and scientists a common set of criteria to describe mental conditions. Entries are often influenced by cultural norms (until 1973, homosexuality was listed as a mental disorder) and political compromise: The manual is written by committees of mental-health professionals who debate, sometimes heatedly, whether to include specific disorders. The entry for GAD, says David Healy, a scholar at the University of Wales College of Medicine and author of the 1998 book *The Antidepressant Era,* was created almost by default: "Floundering somewhat, members of the anxiety disorders subcommittee stumbled on the notion of generalized anxiety disorder," he writes, "and consigned the greater part of the rest of the anxiety disorders to this category."

Critics note that the DSM process has no formal safeguards to prevent researchers with drug-company ties from participating in decisions of interest to their sponsors. The committee that recommended the GAD entry in 1980, for example, was headed by Robert L. Spitzer of the New York State Psychiatric Institute, which has been a leading recipient of industry grants to research drug treatments for anxiety disorders. "It's not so much that the industry is there in some Machiavellian way," says Healy. "But if you spend an awful lot of time with pharmaceutical companies, if you talk on their platforms, if you run clinical trials for them, you can't help but be influenced."

SmithKline's first forays into the anxiety market involved two fairly well known illnesses—panic disorder and obsessive-compulsive disorder. Then, in 1998, the company applied for FDA approval to market Paxil for something called social phobia or "social anxiety disorder" (SAD), a debilitating form of shyness the DSM characterized as "extremely rare."

Obtaining such a new indication is a relatively simple affair. The FDA considers a DSM notation sufficient proof that a disease actually exists and, unlike new drugs, existing pharmaceuticals don't require an exhaustive round of clinical studies. To show that a drug works in treating a new disease, the FDA often accepts in-house corporate studies, even when companies refuse to disclose their data or methodologies to other researchers, as is scientific custom.

With FDA approval for Paxil's new use virtually guaranteed, SmithKline turned to the task of promoting the disease itself. To "position social anxiety disorder as a severe condition," as the trade journal *PR News* put it, the company retained the New York–based public-relations firm Cohn & Wolfe. (Representatives of GlaxoSmithKline and Cohn & Wolfe did not return phone calls.)

By early 1999 the firm had created a slogan, "Imagine Being Allergic to People," and wallpapered bus shelters nationwide with pictures of a dejected-looking man vacantly playing with a teacup. "You blush, sweat, shake—even find it hard to breathe," read the copy. "That's what social anxiety disorder feels like." The posters made no reference to Paxil or SmithKline; instead, they bore the insignia of a group called the Social Anxiety Disorder Coalition and its three nonprofit members, the American Psychiatric Association, the Anxiety Disorders Association of America, and Freedom From Fear.

But the coalition was not a grassroots alliance of patients in search of a cure. It had been cobbled together by SmithKline Beecham, whose PR firm, Cohn & Wolfe, handled all media inquiries on behalf of the group. (Today, callers to the coalition's hot line are greeted by a recording that announces simply, "This program has successfully concluded.")

There were numerous good reasons for SmithKline to keep its handiwork discreet. One was the public's mistrust of pharmaceutical companies; another was the FDA'S advertising regulations. "If you are carrying out a disease-awareness campaign, legally the company doesn't have to list the product risks," notes Mintzes, the University of British Columbia researcher. Because the "Imagine Being Allergic to People" posters did not name a product, they didn't have to mention Paxil's side effects, which can include nausea, decreased appetite, decreased libido, and tremors.

Cohn & Wolfe's strategy did not end with posters. The firm also created a video news release, a radio news release, and a matte release, a bylined article that smaller newspapers often run unedited. Journalists were given a press packet stating that SAD "affects up to 13.3 percent of the population," or 1 in 8 Americans, and is "the third most common psychiatric disorder in the United States, after depression and alcoholism." By contrast, the Diagnostic and Statistical Manual cites studies showing that between 3 and 13 percent of people may suffer the disease at some point in their lives, but that only 2 percent "experience enough impairment or distress to warrant a diagnosis of social phobia."

Cohn & Wolfe also supplied journalists with eloquent patients, helping to "put a face on the disorder," as account executive Holly White told PR News. PR firms often handpick patients to help publicize a disease, offering them media training and sending them on promotional tours. In 1994, for example, drugmakers Upjohn and Solvay funded a traveling art show by Mary Hull, a Californian who suffered from obsessive-compulsive disorder and spoke frequently with journalists about the disorder's toll—as well as her SSRI-aided recovery. Not coincidentally, the companies were awaiting FDA approval to market their SSRI, Luvox, for the treatment of obsessive-compulsive disorder. Among the patients most frequently quoted in stories about social anxiety disorder was a woman named Grace Dailey, who had also appeared in a promotional video produced by Cohn & Wolfe.

Also featured on that video was Jack Gorman, the Columbia University professor who would later make the rounds on Paxil's behalf during the GAD media campaign. Gorman appeared on numerous television shows, including ABC's *Good Morning America.* "It is our hope that patients will now know that they are not alone, that their disease has a name, and it is treatable," he said in a Social Anxiety Disorder Coalition press release.

Dr. Gorman was not a disinterested party in Paxil's promotion. He has served as a paid consultant to at least 13 pharmaceutical firms, including SmithKline Beecham, Eli Lilly, and Pfizer. Another frequent talking head in the SAD campaign, Dr. Murray Stein of the University of California at San Diego, has also served as a SmithKline consultant, and the company funded many of his clinical trials on SAD.

Retaining high-profile academic researchers for promotional purposes is standard practice among drug companies, says Mosher, the former National Institute of Mental Health official. "They are basically paid for going on TV and saying, 'You know, there's this big new problem, and this drug seems to be very helpful.' "

Cohn & Wolfe's full-court press on SAD paid immediate dividends. In the two years preceding Paxil's approval, fewer than 50 stories on social anxiety disorder had appeared in the popular press. In May 1999, the month when the FDA handed down its decision, hundreds of stories about the illness appeared in U.S. publications and television news programs, including *The New York Times, Vogue,* and *Good Morning America.* A few months later, SmithKline launched a series of ads touting Paxil's efficacy in helping SAD sufferers brave dinner parties and public speaking. By the end of last year, Paxil had supplanted Zoloft as the nation's number-two SSRI, and its sales were virtually on par with those of Eli Lilly's Prozac. (Neither Prozac nor Zoloft has an indication for SAD.)

The success of the Cohn & Wolfe campaign didn't escape notice in the industry: Trade journals applauded GlaxoSmithKline for creating "a strong antianxiety position" and assuring a bright future for Paxil. Increasing public awareness of SAD and other disorders, the consulting firm Decision Resources predicated last year, would expand the "anxiety market" to at least $3 billion by 2009. In 2000, the New York chapter of the Public Relations Society of America named the Cohn & Wolfe SAD campaign "Best PR Program of 1999."

THE LESSONS of "Imagine Being Allergic to People" were also not lost on Zoloft's manufacturer, Pfizer. In 1999, Pfizer gained FDA approval to market Zoloft as a treatment for post-traumatic stress disorder (PTSD). Until then, the condition had been associated almost exclusively with combat veterans and victims of violent crime; now, Pfizer set out to convince Americans that PTSD could, in fact, afflict almost anyone.

The company funded the creation of the PTSD Alliance, a group that is staffed by employees of Pfizer's New York public-relations firm, the Chandler Chicco Agency, and operates out of the firm's offices. The Alliance connects journalists with PTSD experts such as Jerilyn Ross, president and CEO of the Anxiety Disorders Association of America, a group that is heavily subsidized by Pfizer as well as GlaxoSmithKline, Eli Lilly, and other drug-industry titans.

In the months following the launch of Pfizer's campaign, media mentions of PTSD skyrocketed. Just weeks after the Alliance's founding in 2000, for example, *The New York Times* ran a story citing Pfizer-supplied statistics on childhood PTSD, according to which 1 in 6 minors who experience the "sudden death of a close friend or relative" will develop the disorder. Other stories highlighted studies promoted by the alliance according to which 1 in 13 Americans will suffer from PTSD at some point in their lives.

Eye-catching figures are integral to disease marketing campaigns, though the quality of the data is sometimes dubious. A report published in February 2002 in the *Archives of General Psychiatry* warned that high estimates on the number of people suffering mental-health conditions often include people whose symptoms are so mild as to not require treatment. "When people look at numbers that say close to 30 percent of the American public has a mental disorder and therefore needs treatment, most would say that is implausibly too high," the study's lead author, William E. Narrow, told the Associated Press.

Many of the statistics used to promote new disorders are taken from studies published in second-tier journals, which frequently depend on direct corporate support. One publication that has drawn fire is the *Journal of Clinical Psychiatry*, whose major funders include GlaxoSmithKline and Eli Lilly. In 1993, the journal published a study claiming that anxiety disorders cost the United States $46.6 billion per year, primarily due to lost productivity. That figure was repeated in countless press releases and made its way into articles in *The Washington Post* and *USA Today*.

The study was produced by the Institute for Behavior and Health, a research firm headed by Dr. Robert DuPont, who served as President Ford's drug czar. The institute's tax returns indicate that its programs are funded almost exclusively by industry research grants; in 1999, for example, it conducted clinical trials on behalf of Merck, Pfizer, and Solvay. DuPont was paid more than $50,000 that year for 10 hours of work per week, in addition to a $56,000 fee that the institute paid to his for-profit consulting firm. The 1993 anxiety study was paid for in part by Upjohn, maker of the SSRI Luvox.

Studies published in medical journals are also useful in reaching a key audience for disease-awareness campaigns—doctors. Physicians, especially general practitioners, are under growing pressure to make quick diagnoses and to treat mental-health conditions with drugs rather than refer patients to psychotherapy. Primary-care physicians now write upward of 60 percent of antidepressant prescriptions, according to the American Psychiatric Association. "There is a pressure to have treatments that are perceived as faster or more efficient," says Dr. Robert Michels, chief of psychiatry at Cornell Medical College.

Drug companies are understandably eager to help physicians identify conditions that can be treated with their products. One widely distributed diagnostic checklist, a 15-minute test that promises to screen for 17 different disorders using special software, was developed by GlaxoSmithKline. Pfizer has funded a test designed to help obstetricians and gynecologists identify women with mental-health problems. According to a 2000 study, sponsored by Pfizer and published in the *American Journal of Obstetrics,* a full 20 percent of all obgyn patients may need psychiatric treatment for anything from depression and anxiety to eating disorders.

Most of all, though, pharmaceutical makers seek to build word of mouth about a condition in the general public—the kind of water-cooler buzz that prompts people to ask their doctor about a disease, and the drug that might treat it. To that end, corporations have increasingly embraced patient organizations that work to publicize mental illness. One such group is the National Mental Health Awareness Campaign, created two years ago to eliminate "the fear and shame that is still strongly associated with mental disorders." The organization is particularly concerned with teenagers, and has run several ads on MTV that encourage unhappy youths to call a toll-free number or visit its Web site. A couple of weeks after the September 11 terrorist attacks, it released the results of a survey, which found that 30 percent of adults questioned felt their mental health had worsened since the tragedy. The group's press release urged "parents and children traumatized by the recent terrorist attacks to avail themselves of the opportunity to speak to mental health professionals."

The campaign's brochures say it has received financial support from the Surgeon General's office. The organization is less forthright about its ties to FoxKiser, a pharmaceutical lobbying firm whose clients include Bristol-Myers Squibb and AstraZeneca. Michael Waitzkin, a partner at FoxKiser, is on the campaign's board of directors, and until recently the campaign was headquartered in FoxKiser's Washington office. (It now operates from the office of the PR firm Health Strategies Consultancy.)

The National Mental Health Awareness Campaign wasn't the only group to step up its profile in the wake of the attacks. On September 26 the PTSD Alliance—the group headquartered in the offices of Pfizer's PR agency, Chandler Chicco—issued a statement warning that post-traumatic stress can affect anyone who has "witnessed a violent act" or experienced "natural disasters or other unexpected, catastrophic, or psychologically distressing events such as the September 11 terrorist attacks." During the following month, according to the trade journal *Psychiatric News,* Pfizer spent $5.6 million advertising the benefits of Zoloft in treating PTSD—25 percent more than it had spent, on average, from January to June.

But the biggest presence in TV drug advertising after September 11 was GlaxoSmithKline, which in October 2001 spent $16 million promoting Paxil—more than it had spent in the first six months of the year combined. In December, the company rolled out a series of new commercials, often broadcast during prime-time news programs and built around lines such as "I'm always thinking something terrible is going to happen" and "It's like a tape in my mind. It just goes over and over and over."

IN THEIR SEARCH FOR new uses, SSRI makers are no longer limiting themselves to disorders with chiefly psychological symptoms. In the March 15, 2002, issue of the *Journal of Clinical Oncology,* Mayo Clinic researchers funded by Eli Lilly reported that Prozac "is a realistic alternative to estrogen replacement for reducing hot flashes" in menopausal women. A recent study at the University of Pennsylvania, funded by the pharmaceutical companies Aventis and Novartis, indicated that SSRIs can decrease the risk of heart attack in smokers.

But by far the most controversial addition to the list of maladies treatable with SSRIs is a condition whose very existence is in dispute: premenstrual dysphoric disorder (PMDD), a female ailment whose symptoms include sharp monthly mood swings and physical pain. PMDD has been listed since 1987 in the Diagnostic and Statistical Manual appendix, which catalogs potential disorders "proposed for further study."

According to Paula J. Caplan, a psychologist and visiting scholar at Brown University who was a member of a DSM committee that evaluated research on PMDD, proponents of including the condition "claimed they were so careful in defining it that it wasn't just going to be someone with cramps during their period. But they were talking about 3 to 5 percent of [menstruating] women. If you do the math as conservatively as possible, 3 to 5 percent gives you one and a half million women [in the United States]." Caplan resigned from the committee before it voted to list PMDD in the appendix.

Though the condition remains controversial in the medical profession—one 1992 study found that men and women suffered from PMDD's symptoms at almost the same rate—its inclusion in the DSM proved a godsend for Eli Lilly, the manufacturer of Prozac. In 2000, the company gained FDA approval to market Prozac as a treatment for the condition; Eli Lilly promptly repackaged Prozac as a pink-coated pill called Sarafem and launched a PR campaign warning that "millions of menstruating women" suffer from PMDD. "Does juggling work, family and personal commitments leave you feeling frazzled and stressed out?" the Sarafem Web site asks. "We have some tools to help."

The idea of characterizing uncomfortable menstrual symptoms as a mental disorder troubles Caplan, who wonders where the medical community will draw the line. "I could say to you, 'Well, your propensity to call people and ask them probing questions is a disorder,' " she says. " 'We'll call it intrusive exploratory disorder.' "

No such malady is yet listed in the DSM. But the quest for new uses for the SSRIs is continuing. At last year's annual convention of the American Psychiatric Association, researchers presented a major study on a new "hidden epidemic"—compulsive shopping. Jack Gorman, the Columbia psychiatrist who had earlier helped publicize anxiety disorders, made another appearance on *Good Morning America* to discuss the new condition, which host Charles Gibson told viewers could affect as many as 20 million Americans, 90 percent of them women. In the wake of the new study, Gorman said, scientists would "almost certainly" look into treating the disease with SSRIs.

The study in question was funded by Forest Laboratories, for which Gorman has served as a consultant. A laggard in the SSRI business, the company hopes to carve out the compulsive-shopping niche for its pill, Celexa. Expect the publicity machine for something akin to "persistent purchasing disorder" to rev up soon.

JOSEPH D'AGNESE

An Embarrassment of Chimpanzees

FROM *DISCOVER*

Along with the ethical quandaries posed by medical research on animals, there is another issue that is not as widely discussed: What happens to these animals when they are no longer needed by laboratories? In the case of chimpanzees—humankind's closest cousin—the problem is severe. There are more than sixteen hundred research chimps in the United States, but precious few establishments that can serve as sanctuaries for them; resources are limited, and many communities object to having animals with infectious diseases in their midst. The journalist Joseph D'Agnese makes a heartbreaking visit to the Fauna Foundation, outside Montreal, where chimps play cards, watch television, and live out their last years far removed from their past traumas.

I first heard the story a year ago. I was interviewing a scientist when he began griping about how difficult it is to get a chimpanzee for medical research. "They're expensive," he said, "and you've got to pay all this money into a social security plan to take care of them when they retire." Retired chimps? Just where do they go to retire—and what do they do when they get there? Eat bananas? Play shuffleboard? "I heard they put them on an island sanctuary in Liberia," the researcher said. He didn't know much more than that, and of course it wasn't his job to know. He is the scientist. He uses chimps to answer scientific questions. Someone else deals with what comes afterward.

For a while I thought about going to Africa to look for that chimp island. In my mind's eye I envisioned a paradise where repatriated American chimps lived free of humans, free of the cages that once confined them. But that turned out to be a pleasant fiction, a tale told to lab workers foolish enough to ask. By then I had observed chimps in zoos, read about them in the scientific literature, and immersed myself in a world of animal sanctuaries that is stranger, more interesting, and more disturbing than I could have imagined.

Eighteen chimps do live on a pair of islands in Liberia, most of which were bred on-site by an American hepatitis research laboratory. And sanctuaries throughout Africa shield wild chimps from poachers. None of these places will accept U.S.-bred lab chimps after we're done with them. That's why sanctuaries are needed here. At this moment, the United States is up to its ears in chimps. During the 1980s, laboratory supply companies bred chimps like crazy to meet the demands of AIDS and hepatitis researchers. That didn't work out too well. By the late 1990s researchers conceded that while some chimps become HIV-positive, almost none develop full-blown AIDS. At least 200 chimps have been exposed to HIV, yet only two may have died of AIDS. The researchers switched to macaque monkeys. For a short time the National Institutes of Health, which funds much of the biomedical research in this country, considered killing HIV-exposed chimps when they were no longer useful. The NIH later decided not to, in part because the animals are listed as an endangered species. But the surplus has mounted—today more than 1,600 live in various primate facilities in the United States—and humans have begun to ask themselves a serious question: What are we going to do with these animals?

During his last weeks in office, President Clinton signed the Chimpanzee Health Improvement, Maintenance, and Protection (CHIMP) Act, which mandates a national system of sanctuaries for chimps who qualify, but it is likely to be two years before any new refuges are ready. In the meantime many animals will have to remain in labs. So far, about 200 chimps have been earmarked for retirement. When chimps do enter sanctuaries, there will be a string attached: If a sanctuary owner takes government money, he or she must be prepared to send the chimps back to the lab for further research if asked to do so—a stipulation that infuriates those who believe that plucking a chimp out of retirement negates the concept of sanctuary.

IN 1997 New York University decided to get out of the chimp business and shut down its Laboratory for Experimental Medicine and Surgery in Primates. It's a common story. Around the world, chimp labs are dwindling: The last remaining facility in Europe—in the Netherlands, with 105 chimps—is closing. New Zealand has banned chimp research, and in the United Kingdom no new licenses are being granted for this work. The only other nations that still use chimps for medical research are Japan (370 chimps), Liberia (18 chimps), and Gabon (72 chimps).

When labs close, chimps are up for grabs. Of the 250 chimps who contributed, as they say in lab parlance, to experiments at the New York University lab, 90 were placed in sanctuaries; the rest were transferred to other labs. The most difficult to place were those that had been exposed to HIV or hepatitis, both transmissible to humans.

Later in 1997 the Fauna Foundation outside Montreal became the first sanctuary in North America to give retired, HIV-exposed chimps a home. Fauna had been an animal refuge for close to 10 years, and the chimps joined a motley crowd of goats, pigs, chickens, rabbits, horses, turkeys, geese and ducks, cats and dogs, Scottish Highland steers, cows, llamas, emus, rheas, capuchin monkeys, one guanaco, one Jacob's sheep, one ostrich, and one donkey. Until Gloria Grow, 46, the owner of the sanctuary, and her husband, Richard Allan, 49, announced their intention to shelter 15 of the chimps from the New York University lab, eight of them HIV-infected, the refuge had been regarded by residents of Chambly as a harmless oddity. Suddenly, the local planning board challenged every variance Fauna requested. When the board saw plans for an elaborate, secure chimp house complete with cages, they acquiesced. Nonetheless, teachers conducted chimp drills at the elementary school, instructing kids to hide in the classroom closet if a chimp appeared in their midst; and police laid in a store of Tyvek suits and tranquilizer guns.

Fauna's chimps live in a 9,000-square-foot building that looks a bit like a day care center. Despite the cages, living conditions beat lab life knuckles down. The outdoor play area contains picnic tables, chairs, and swings; indoors, two-story playrooms are packed with toys, blankets, and more swings. Chimps can also rest in private cages that give them access to the indoor play space but keep them separated from humans. They can snack on fresh fruit and vegetables, or page languidly through Victoria's Secret catalogs. The human form enthralls great apes.

Each day Grow and her staff of three whip up three savory meals. The menu includes fruit, oatmeal, spaghetti, potatoes, soups, stews, steamed vegetables, and rice, as well as the occasional vegetarian pizza and birthday cake. Staffers mix gallons of concentrated orange juice every day and laboriously pour it into empty water bottles with plastic lids. "Most of the chimps know how to unscrew them," an employee explains, "but sometimes they like to stick a canine in the cap to bite it off."

One day last October, the chimps are lounging around after lunch, picking at their food, grooming themselves, and playing with toys. Some snatch plastic cups of hot Tetley tea off the trolleys parked in front of their enclosure, sip carefully, and return the cups through the bars without spilling a drop. Grow brushes Tom's back with a small brush. "Let me see your fingernails," she says. Tom holds them up for inspection. Another chimp, 42-year-old Annie, the oldest and a surrogate mother to the others, spots the brush and gestures through the cage for it. "You want the brush?" Grow asks as she slips it to her. Annie spends a couple of blissful minutes stroking her coat.

A few minutes spent watching chimps manipulate objects like cups, bottles, and brushes quickly demonstrates why biologists regard them as the top tool users in the animal kingdom, after us. Besides being dexterous, they are intelligent, strong, and often aggressive, especially as they grow older. Chimps also seem to have a sense of humor, which any visitor to Fauna notices immediately: They delight in teasing humans as well as each other. They routinely spit water at their caretakers, cleverly varying the pattern to confuse them. They also seem to understand and respect social hierarchies: A beta male accepts his lot when an alpha male swipes his orange but goes ballistic when a lower-ranking female does the same.

The chimpanzee's ability to learn can be humbling. In 1967 psychologist Roger Fouts taught chimps to use American Sign Language, which they mastered and taught to other chimps. Since 1983 psychologist Sarah Boysen has been teaching chimps at Ohio State University to do simple arithmetic; in 1991 she figured out how to teach them fractions. In 1999 a landmark paper written by Jane Goodall and eight other prominent primatologists established that chimps use their smarts to master their environment. Chimps can codify cultural behavior—how to hunt, how to eat ants, how to groom oneself and others—and pass that knowledge along to their young ones. Chimps that live in the Gombe forests of Tanzania have been observed dancing, apparently to make the rain stop.

In the wild, bands of chimps will rove the jungle for six or seven miles a day, joining together to hunt monkeys, which they eat with relish, usually after dashing out the smaller creatures' brains. That's the side of chimps humans rarely see or choose not to see. However, they can also be kind. A chimp was once observed trying to help a wounded bird to fly at a zoo in England.

At Fauna, there is constant physical contact between humans and chimps.

The tiniest scrape or scab on Grow's hand will elicit concern from a chimp, usually in the form of a kiss. According to the standards of the Centers for Disease Control, Fauna's chimp house is a biohazard facility. If this were a U.S. lab, workers would be required to wear Tyvek suits, goggles, masks, or hair nets, mandatory garb worn by researchers studying HIV or hepatitis. But Grow and her workers wear street clothes, unless a chimp has an open wound or needs surgery, in which case they follow aseptic procedures, donning gowns and gloves just as they would with a human patient. They believe that if the animals are treated well, they will not harm their caretakers. The theory has proved true so far, but the chimps do get into fights with each other that require bandages or surgery.

Now Grow calls to another animal: "Billy Jo, is your show on?" She peeks at the TV. "Oh, it's *Rosie.* Don't worry, *Oprah* will be on soon." Grow agonizes about keeping the chimps stimulated. Because they remain caged, she wants to help entertain their restless spirits. Hence the painstaking preparation of meals, classical music piped over the stereo, hanging spider plants, brushes and paints, Halloween decorations, Christmas lights, birthday parties, crackling fires in the wood-burning stove, strands of red licorice, and aromatherapy candles.

In the United States, lab animals fall under the jurisdiction of the Department of Agriculture, which inspects labs and enforces the Animal Welfare Act of 1985. By these standards, the Fauna chimps were treated well in their former lives: They got adequate food and shelter, their cages were clean, and they received the occasional toy or orange. But Grow and others like her consider those standards weak and seek to do better. "I want them to be happy," Grow says. "To treat chimps well, you should treat them as you would victimized people. Because they have been victimized. Terribly. Oh, it's so awful what they've been through."

Annie, for example, was born in Africa, probably in 1959, then captured and sent to the United States. She gave her life to humans for more than 35 years— at least 15 in the circus, followed by 21 in the lab as a breeder. When she refused to mate, she was artificially inseminated. Her child was transferred to another facility at age 3. Another Fauna chimp, Rachel, was born at the Institute for Primate Studies in Norman, Oklahoma, in 1982. Rachel was sold for $10,000 as a pet but ended up at the New York University lab when her owners divorced. Rachel, who had grown up taking bubble baths and prancing around in dresses, spent the next 11 years isolated in a cage. Today she occasionally bursts into screaming and scratching fits, lashing out at her own hand, apparently because she thinks it is attacking her. Her outbursts have diminished somewhat since she arrived at Fauna in 1997, but her body is still covered with self-inflicted sores.

"Jeannie was going to be euthanized—did you meet her?" Grow asks. Lab workers had to medicate Jeannie, an HIV-exposed chimp, to stop her seizures, during which she ripped out her fingernails and thrashed any human or chimp nearby. "She had a nervous breakdown before she came here, but she has made lots of strides in her development. They all have. They put on weight, grow more hair, their coats are shinier. They sleep better at night. We don't have nearly as many fights as we used to when they first came, and they have learned to vocalize more like real chimps."

Fauna's annual budget for the entire farm is $60,000. The food bill is $40,000; the rest covers medicine and necessities such as bedding straw, tools, and equipment repairs. In a good year, $15,000 of that comes from private donations. Fauna is not eligible for funding under the CHIMP Act because the sanctuary is in Canada. Grow says she would not apply for funds even if Fauna were eligible because of the requirement that sanctuary owners return chimps to labs on demand: "I would never send them back. Who would?" The bulk of Fauna's operating revenue comes from a dog grooming business and from Allan's veterinary clinic. On the first night of my visit, Allan, a French-Canadian who has been doctoring animals on the outskirts of Montreal for 27 years, arrived at dinner in scrubs, looking exhausted. Delighted that Fauna would be the subject of this article, he quipped, "Tell them we need money."

ONE RECENT COLD and wet morning, Grow is chopping fresh vegetables for the rabbits and pigs when her sister Dawna Smith, who works in the chimp house, chugs up to the barn in a Volkswagen. "Get in," she calls to Grow. "I need you to come look at Pablo."

"What's wrong with him? He was fine last night."

"In."

Up at the chimp house, the 30-year-old, almost 200-pound chimp struggles to make himself comfortable in his nest—a pile of blankets on a 12-foot-high platform inside the chimp house. He can find no peace. First he sits, then he stands, repeating the process over and over: sitting, standing, sitting, standing. He wheezes constantly.

Since the day of his arrival five years ago, Pablo has been ill. One winter he developed a cough that X rays showed to be bronchitis. Medication helped, but each autumn Grow worries that Pablo's cough will return. Still, she has never seen him behave like this. She dashes up a spiral staircase to offer him more blankets, an antibiotic, and a Tylenol. The big-lipped chimp graciously accepts the blankets but spits out the pills. Grow runs out to find her husband, who is busy spreading a load of red cedar mulch someone has just donated to the farm. Allan's practice, largely cats and dogs, did not prepare him for the variety of animals with which he now shares his land and board. To prepare for the chimps' arrival, he spent a few days training at the New York University lab with veterinarian James Mahoney.

"What do you think is wrong with him?" Grow asks.

"He's dying," Allan says, staring into the cage.

Grow doesn't want to hear that. Her husband often acts on the premise that there is only so much one can do for an animal, especially a wild animal who will not permit a detailed physical exam. But Grow was raised to believe that she should go to extremes to help sick animals. Her father, an electrician, thought nothing of stopping his truck in heavy traffic to rescue a wounded seagull. Now, as Grow looks on, Allan phones Mahoney and leaves a message on his voice mail, then heads off to resume mulching.

Hours tick by. At lunch, Grow and her sister, employees, and volunteers sip soup and munch eggplant casserole in silence. When Allan comes in to wash up, Grow asks, "What do you think we should do?"

"What should we do?" Allan repeats. "We should wait and see how he feels tomorrow."

"Wait and see? If I were one of your patients, you think I'd want to hear that?"

"What do you want to do?" says Allan. "Tranquilize him?"

When Allan heads back outside after lunch, Grow asks him again what he thinks is wrong. He repeats the two words he uttered that morning. The words stab the air, and then he is gone. Grow is left to ponder their meaning in a congress of women. Pablo can't be deathly ill, she and her staffers decide. He is only 30; captive chimps can live to be 60. Her sister dissents. "The thing is," Smith says in a measured tone, "Richard's always right."

Mahoney phones at 2 P.M. The big chimp is down, still breathing hard. He drank some juice with antibiotics but vomited it up. Mahoney offers possible diagnoses: pneumonia, a cardiac problem, a twisted intestine. Given Pablo's past, pneumonia seems likely. Allan is instructed to administer three injections, one after the other: an antibiotic, a diuretic, and cortisone for shock. If Pablo has pneumonia he should feel better after the first shot. Allan drops the phone and dashes to get his bag.

In labs, monkeys and chimps are trained to present their arms for blood draws. Pablo had always resisted, so he was usually tranquilized—"knocked down," as the lab techs say, with a dart fired from an air pistol. Allan fears he'll have to break out the darts for the first time ever at Fauna. When Pablo sees the needle, he thrusts his arm out. Allan is stunned. "This guy never liked needles, but he gave me his arm. Didn't put up a fight."

Minutes after the injections, Pablo lies back and closes his eyes. His face is immobile; a black arm hangs limply off the side of his nest. Allan carefully unlocks the gate to the chimp enclosure, and Grow rushes up a ladder. She grabs Pablo's hand and feels a twitch. Life shudders out of the great ape's body. She begins to cry but manages to help carry the body to the floor. Allan confirms he'd dead, and Grow insists that the humans leave the enclosure to allow the other chimps a chance to see Pablo.

Normally, when a lab chimp dies, he dies alone in a cage and is whisked away. Grow believes chimps should be allowed to witness everything. A couple of times Allan has performed surgery in the kitchen, where all the chimps could see him. "Someday, when I die," Grow says, "I want to be placed right here where they can all see me and know that I am gone."

So, as Grow and her staff sit weeping outside the enclosure, the chimps approach Pablo. Alone or in pairs, they tug at his arms, open his eyes, groom him, rub his swollen belly. Annie pours a cup of juice in his ear. Grow says it might be an attempt to annoy Pablo and wake him up. Before long, the chimps wander off, hooting. The hoots blossom into screams, and soon the walls of the chimp house echo with the sound of knuckles pounding steel.

THE NIGHT OF Pablo's death, Allan conducts a hasty necropsy, but neither he nor his colleagues have had much experience handling a large and potentially infectious animal. His veterinary clinic is well-supplied with Tyvek suits and latex gloves but short on masks and goggles. Everything seems too small for Pablo's frame: the clinic's back door, the operating table, especially the freezer into which Allan and a sobbing Grow stuff his body when the procedure is finished.

Eighteen days later, after Grow has pleaded unsuccessfully with different agencies to perform an official necropsy, the Montreal health department presses a pathologist into service at the vet school in Saint-Hyacinthe. The immediate cause of death is listed as an acute lung infection, but the physician who examined the body also found an abdominal infection and mild hepatitis. Internally, the animal's organs were crisscrossed with thick, fibrous scars, most likely the remainders of various procedures. To do an animal biopsy, a technician uses a punch to clip out a chunk of tissue. The procedure leaves a large hole that, if infected, can take years to heal. Pablo was also vulnerable to infection on another front. Darts fired from an air pistol are, by definition, non-sterile; each penetration carries germs from the surface of an animal's skin into its body.

According to his research dossier, Pablo, known as Ch-377 at the New York University lab, had been darted 220 times, once accidentally in the lip. He had been subjected to 28 liver, two bone-marrow, and two lymph-node biopsies. His body was injected four times with test vaccines, one of them known to be a hepatitis vaccine. In 1993 he was injected with 10,000 times the lethal dose of HIV. The barrel-chested chimp had shrugged off AIDS and kept hepatitis at bay only to die of an infection aggravated by years of darts, needles, and biopsies.

"We always knew the chimps had a lot of problems," Grow said two months after Pablo's death. "But we always thought they were problems we could take care of—because they were on the outside. Now we are learning that there are a lot of things going on inside them that we may never know about. Annie's sick now. Jeannie's sick now. What happened to Pablo wasn't unusual; it was average."

Activists insist that animal-free science is already here—in the form of in vitro research, data gleaned from autopsies, clinical observation, and epidemiology. But the scientists who work with chimps say that the inoculations, biopsies, and knockdowns, though regrettable, are necessary. "I think the idea of moving to humans is nonsense," says Alfred Prince, the hematologist who heads chimpanzee research in Liberia. "Ethics committees in hospitals are getting tougher and tougher, and the work you can do in people is less and less. We will probably always need animal models . . . I think the answer is, if you're going to do this work on chimps, you better take really good care of them." Other researchers, including primatologist Roger Fouts, believe that the days when we are willing to imperil an endangered species for our own sake may be numbered. Until then research will proceed, and people like Gloria Grow will be left to deal with the results, as she did in January 2002, when Annie, the grande dame of the chimp house, died. Her body is awaiting a necropsy. Then the body will be sent, like Pablo's, to a local crematory that donates its services to Fauna. Grow plans to bury some of the ashes of both animals at the sanctuary. Sometime soon, Jane Goodall will take the rest of the ashes with her to Tanzania to sprinkle in the forests of Gombe, where chimps dance to stop the rain.

DANIELLE OFRI

Common Ground

FROM *TIKKUN*

Doctors routinely go to great lengths to care for their patients but, beyond a dose of bedside manner, maintain a professional distance from their patients' personal lives. Doing so is not always easy, however. Danielle Ofri, a doctor and writer, recollects an episode from the early days of her career when her conscience, and her own past experience, prompted her to reach out personally to a patient.

"We are a Catholic medical center, Dr. Ofri." The medical director leaned back in his chair across from my desk. "Do you have any issues with that?"

His gray hair was severely parted on the right and I could trace the individual strands that were tethered down on the side by hair grease. A stethoscope peeked out of the pocket of his tailored blue suit. He had just finished his long introductory speech with me, enumerating the vast array of services and the selling points of his medical group. He was clearly trying to impress me with his institution. After all, the reason I was doing a temp assignment here was because they were short-handed and looking to hire.

I was caught off balance by the question. What could he be driving at? Was my Jewish background an issue here? Was my last name too "ethnic"? I paused and then slowly asked back, "Should I have issues?"

"Well," he replied, in his careful New England lilt, "we do not promote birth control. If a patient requests it, we will provide it. But we do not offer it, promote it, or condone it."

Before my super-ego could grab control, my New York sassiness spilled out. "So, I don't suppose you perform abortions, do you?"

I could not believe I had just said that.

The older physician did not appear fazed. "No, we do not terminate pregnancies. Nor do we permit referrals to physicians who do. If a patient requests that service, we have them call their own insurance company. Their insurance companies make the referral."

He stood up and put out his hand. "We are glad to have you aboard, Dr. Ofri. We hope you enjoy your six weeks with us. And," he paused with a smile, "we hope you consider staying longer."

I remained in my office after he left, a little confused about what I had just heard and very embarrassed about the sauciness of my retort. I finally brushed it off, attributing it to high-level politics that I was not a part of.

I had never spent much time in New England before. The town looked just as I had imagined. Regal Victorian mansions with wrap-around wooden porches lined the main street. Well-tended rose bushes graced the picket fences. Manicured shrubbery lined the driveways. A river meandered through the town and I often saw kayakers as I drove over the small bridges each morning in my beige rental car. This was a different planet from my native New York City.

I had been assigned to a small private practice that was short-handed after two doctors had moved away. The staff members welcomed me warmly. They gave me a large office with three exam rooms in a separate wing of the suite, and a nurse, Karen, to work exclusively with me. At the beginning of each appointment Karen would take a brief history from the patient, check their vital signs and jot down their medications. When I entered into the room afterward to see the patient, I would find all the supplies that I might need for that particular patient neatly laid out. I learned that the walls of the examining rooms were fairly thin because when I was finished with the patient, Karen would be waiting outside with whatever vaccines or medications I had discussed with the patient.

This was nothing like Bellevue Hospital—the city hospital where I did my residency. Practicing medicine had never been so easy! I noticed that the medicine cabinet was stocked with free samples of birth control pills along with the anti-hypertensives and cholesterol medications. Apparently, no one took the contraception rule too seriously.

Nobody ever bothered Karen and me in our little corner. It was as though we had our own practice. Between patients we would share stories of her life in New England and my experiences at Bellevue. And I loved that she kept a picture of her golden retriever, Sam, on her desk.

Three weeks into my assignment I met Diana Makower, a young computer programmer at a local financial firm. She was wearing a gray suit with a purple silk blouse. A single strand of pearls hung around her neck. Her carefully applied make-up had started to smudge from the tears slipping down her cheeks. "I think I'm pregnant," she spilled out, almost before I could introduce myself. "I did one of those home pregnancy tests and it was positive. All I need from you is a blood test."

I put down my stethoscope and pulled up a chair.

"It's a complicated situation," she wept. "I am ending a relationship with my boyfriend, but it wasn't him. I have an old friend, it's never been more than that, but I think he and I might be developing a romantic relationship. We slept together just once, three weeks ago. I really think we could have a serious relationship, but it is not ready for this. I can't believe this is happening."

"If you do turn out to be pregnant," I asked, "what do you think you would do?"

"I need to have an abortion. I can't have a kid now; I'm single, I don't have a stable relationship yet. I'm not ready for it now."

"Are you sure that's what you want to do? Have you considered other options, like adoption?"

"Absolutely," she said. "I have made my decision. I just need to know where to go."

I suddenly thought of the medical director with his slicked-down gray hair. According to the rules, I was supposed to tell Diana to call her insurance company. Her insurance company? I had visions of a bored bureaucrat slurping on his coffee while dispensing advice on a delicate matter to my distraught patient. How could I send Diana into a situation like that? I excused myself and went to consult Karen.

Karen did not know which local doctors performed abortions. "I stay out of that mess," she said. The Catholic hospital that the practice was affiliated with certainly did not. She sympathized with my predicament but warned me not to let the office manager know what I was doing. "Someone else gave out a phone number once," she said, "just a phone number. It wasn't even documented in the chart, but somehow it got out and they got into trouble."

I stared out the window and could see my rental car parked in front of a clapboard house across the street. The house was painted bright yellow with pale blue trim. A wooden porch surrounded three sides of the house. It was overflowing with hanging spider plants and overripe ferns. Wicker furniture with floral cushions was arranged around a wrought-iron table. An American flag dangled from a second-story window. This Catholic medical institution might choose not to perform abortions, but what about my ethical duty to provide the care my patient needed? Sending a distressed patient to an 800 telephone number would not hold water under the Hippocratic oath.

It seemed clear to me that my duty was first to my patient, and only secondarily to some faceless institution. Unfortunately, as a stranger to this small town, I did not know the local resources. I didn't know the names of the nearby physicians to even make the referral if I had wanted to break the rules. I suddenly pined for Bellevue, where I knew all the doctors and I knew the system. If I needed help, all I had to do was dial the operator and have the appropriate doctor paged. I looked back at the yellow and blue house across the road. It seemed hostile and antagonistic. The small-town civility made me feel claustrophobic.

Grinding my teeth, I re-entered the exam room. "As you may know, this medical practice is Catholic," I told Diana, "so we cannot provide referrals for abortion. The truth is, I wouldn't know where to send you even if I could. The rule is that you are supposed to call your insurance company and get the referral yourself. I would do it for you, but I can't. However, if you get the list of possible referrals, I will call around to find out which is the best."

Diana nodded, and then asked if she could be alone. I left her with a box of tissues and told her she could stay as long as she liked.

I called Diana the next day to let her know that the repeat pregnancy test was positive. When I called, I got her voice mail at work. She had told me that it was a private line, but suddenly I felt paranoid. I did not indicate that I was a physician and I left a cryptic message about results being "confirmatory of our original data."

Diana returned my call a few hours later. Her insurance company had given her two phone numbers, without names, in the next state over. Her health plan had no gynecologists in this state who performed abortions. Nobody in the state? My patient couldn't get the care that she needed in her home state? I was horrified. How could I send her off into the unknown like that? How could I abandon her to a couple of random, blank telephone numbers in another state? I felt like we were back in the 1950s, sneaking around with code words, no names mentioned, having to go out of state for an abortion.

I plowed through my roster of patients for the day, but I couldn't focus on the coughs, rashes, and shoulder pains. All I could think about was Diana. I imagined her driving over the state line, tears pressing at her lid margins. The lonesomeness in the car, the bitter highway, the directions scribbled on the back of a used envelope. I imagined her squinting at the scrawled directions, the car slipping ever so slightly out of the lane as her mind diffused focus from the highway median to the second left after the traffic light to the enormity of what lay ahead. Then she would tighten her grip and the car would even out. She'd admonish herself to watch the highway. And so she would watch the highway, look at the highway, stare at the highway, until the yellow lines would begin to quiver, then shudder, then melt into the saltiness dribbling down her face.

Between patients I paced around my office, too irritated to sit still. What kind of place was this where some administrative rule could interfere with patient care? Wasn't patient care more important than a bunch of rules? I wondered when was the last time any of those bureaucrats had actually seen a patient. When was the last time they'd sat face-to-face with a patient, watching the tension lines around the mouth tremble, smelling the moist desperation, accepting the burden and the honor of tender secrets? I fumed all afternoon, cursing the insurance companies and the politicians whose ideologies and business concerns were elbowing into my office, into the sacred space that my patient and I shared.

Then Karen told me that the wife of one of the doctors used to work at a teen clinic. Grateful for this information, I called immediately. She knew of those two out-of-state facilities and told me they had reputations for treating patients like cattle. There was, however, a private women's clinic two hours north that was professional and reliable. But most insurance companies would not cover the cost of the procedure.

I called Diana at home that evening. She had already made an appointment at one of the out-of-state clinics and was very appreciative of my "insider information." I gave her the number of the private women's clinic.

"Have you told him?" I asked.

"No. No, I can't tell him. Not yet, at least. Maybe afterward."

"Is there anyone that you'd feel comfortable talking to, a friend, a family member? Is there someone who could come with you?"

"No, not really," she replied. "I mean I have good friends, but I couldn't tell them about this. They wouldn't understand."

I winced at the thought of her going alone. There was a sense of something shameful, something to hide. "Bring your own bathrobe," I added, before we hung up. "It's more comfortable than a hospital gown."

I called her again the following day. Just to make sure she was okay. We chatted a bit and it turned out that she had grown up in New York.

"Really?" I asked, excited to uncover a fellow New York native here in the wilds of New England. "Where were you born?"

"Queens," she said, "but then we moved out to Long Island, which is where I really grew up."

"My family did something similar. I was born in Manhattan but then we moved out of the city to Rockland. I hated the suburbs, though. I never forgave my parents for leaving the city."

"Me too," Diana said. "I spent all of my high school years hanging out in the city, trying to make up for my parents' foolish flight to the 'burbs. My friends and I would take the train in on weekends and hang out in Greenwich Village."

"So did I," I said excitedly. "We used to tramp up and down Bleecker Street then go hear music at Kenny's Castaways."

"I know Kenny's Castaways. The club that never checked ID."

"That's the one. Kenny's Castaways. And you went to Le Figaro Cafe, didn't you?"

"Absolutely—southwest corner of Bleecker and MacDougal. That's where I had my first cappuccino. I couldn't bear to drink my parents' instant coffee after that."

I left work that evening and drove to my hotel. The very act of driving, of commuting by car, made me feel odd. It had been more than fifteen years since I'd relied on a car for transportation. In New York I was a regular denizen of the subway, and an avid bicyclist. I particularly relished gridlock traffic in Manhattan. I adored watching the irate drivers fume inside their cars, locked in the daily midtown mess, while I whizzed past on my ratty old ten-speed, needling my way in and out of unloading trucks, yellow cabs, and wayward pedestrians.

And now as I sat in my rental car, idling at a traffic light, I felt confined. I pined for the freedom of my bike. I yearned for the foot-based culture of New York, in which everything I needed was in walking or biking distance.

Some people feel nervous in big cities; I feel nervous in small towns. No pedestrians on the streets. No one to make eye contact with. No one to negotiate personal space on a sidewalk with. No mass of actual human beings on the street to remind you that you are alive and part of a species. Only cars.

And so I sat in my car, cut off from humanity, isolated in a metal box that rumbled with diesel heat under my feet as the traffic light languished on red. Sure, the old houses were beautiful to look at and the landscaping impressive, but there were no people. I craved people. Stuck in my car, I could think only about Diana. She was also cut off. There was no one she could confide in, no one she could bring with her. I realized that I was probably the only person in this world she had spoken to about this. In the small enclosed space of my car, with that bland smell of whatever they use to make seat stuffing, the heaviness of that burden weighed onto the cramped muscles of my shoulders. There were hundreds of people tucked into similar steel automobiles who were riding along the same street as me—hundreds of cars shuttling human beings within their tiny isolated orbits—but there was only one that contained Diana Makower's confidence. As a woman, I felt an almost sisterly duty to be there for her during this uniquely feminine quandary. As her doctor I felt that I had the responsibility to make sure she got the medical care she needed and felt guilty that I couldn't help her more directly. And as a human being, as the driver of the steel box that held her confidence, I felt the moral obligation to hold that dear, to treat that confidence with the utmost respect. I couldn't abandon her during this difficult and lonely period.

When I arrived back at my hotel, I called her again. Just to see how she was doing. Two days later I called Diana again. I somehow found a pretext to call her almost every day until her abortion date the following week.

I felt a bit more like a therapist than a physician and I understood why therapists are to keep their personal lives out of the therapy. Therapy is about the patient, not about the therapist.

I ached to share my own experience, but professionalism, and I suppose some lingering shame, prevented me. I'd been only seventeen at the time and just returning home from my first year in college. I had passed my calculus final exam and was pretty sure about physics. I had turned in my last organic chemistry lab report. I was about to go off to be a counselor at summer camp when I discovered that I was pregnant.

I'd had a steady boyfriend the entire year. Before we got involved I had gone to Planned Parenthood because I didn't want to be irresponsible. I remembered the long talk with the counselor in the windowless room with the overly cheery posters. We'd decided together on the diaphragm for birth control. The package insert listed a 95 percent effectiveness rate. No one ever spoke about the other 5 percent.

I lived in New York, the most liberal city in the most liberal state. My friends and parents were all liberal, pro-choice people. But I was too scared to tell anyone; it just didn't seem possible that it was happening and it didn't seem possible to tell anyone.

After the pregnancy test I sat in a park and cried alone. It was a park where my family used to have picnics when I was little. My parents would buy a pre-cooked chicken from the nearby kosher deli. We'd bring paper plates and the vegetable salad. And of course, our dog Kushi. This was her chance to run off the leash. Sitting in that park now I longed for the smell of her soft black fur. I craved her warm, all-accepting dogness to snuggle up to. Someone to whom I wouldn't have to explain all the complicated human confusions. But she'd died the previous year, just before I'd left for college.

I arranged an appointment at a local women's clinic. That night I made a long-distance call to my boyfriend. The geographical and personal gaps were apparently too vast to bridge—he couldn't quite accept what I was telling him over the phone. And he didn't offer to help me pay for it.

The next day I lied to my parents about having a party to go to so I could borrow the car. The clinic had said to bring a comfortable bathrobe. I snuck my mother's out of her closet.

The drive was eerily dissociated. The yellow lines in the road didn't seem parallel to the outer curbs. They listed and buckled, slighting the rules of Cartesian geometry. They drifted to other planes, to the odd dimensions of irrational numbers. Then they'd swing back with a jolt, clobbering into my focus. As the car shuffled closer and closer to the clinic, I felt my body shrinking. It dwindled within itself until there was nothing left but a little girl who desperately wanted her dog.

I lugged myself, or what little was left of myself, up the steps. I registered a name—I think it was mine—and followed the nurse into the back. She instructed me to change into my bathrobe and wait in the main room until I was called. The room was filled with eight or so women in different-colored bathrobes. We could have been at a slumber party, except that no one was smiling. Some magazines were scattered on the table, but the articles were about beef casseroles and electricity-saving tips. I pulled my mother's flannel robe around me and concentrated on the orange industrial carpeting. It really was orange, although if you looked carefully, there were lonely bits of red and yellow scattered within.

They gave me a choice of general or local anaesthesia. The budding college-educated scientist wanted local, wanted to know everything that was going on, wanted to control the whole biology experiment. But the little girl who yearned for her dog immediately chose general. I didn't want to know. I didn't want to remember.

I awoke crying in another room. It was overly bright and the sheets were stiff. My stomach pulsed with an alien ache. The nurse said to stop acting like a baby, it didn't really hurt that much. I checked out and went back to the same park to cry some more.

A week later, a letter arrived from my boyfriend. He told me that he felt terribly guilty. As "penance" for himself, he said he could never be with me again. That summer was long and lonely.

In the years that have gone by I have told almost no one. Part of me feels that I should be contributing to the destigmatization of abortion by being open about my own experience. Yet another part of me feels it is something personal. Worse yet, something to hide. I feel guilty and hypocritical.

Sometimes I think about the child that might have been. At seventeen, I had precious few resources to raise a child. I would never have finished college, much less gone to medical school. I might have faced a lifetime of minimum-wage jobs and food stamps. What would my child's life have been like?

I called Diana after her abortion. She told me that the staff members at the clinic were extremely kind and supportive, and that it didn't hurt too much. I breathed a sigh of relief. We spoke a few more times after that. Each time I felt the urge to share my story, but I couldn't.

I am not a politically active person. So much of what transpires in the government seems to have no bearing on my life; I just want to take care of my patients and my family. The decision about abortion is a difficult one, not one that I would wish anyone to face. But when I see teenage mothers in my clinic with minimal education, no job skills, barely mature enough to take care of themselves let alone the two or three babies on their laps, I am viscerally aware that my life was at the mercy of laws that permitted access to safe abortion. A different time or a different place and the outcome could have been vastly different.

Doctors often unconsciously separate themselves from patients—they are the sick ones and we, in our white coats, are different from them. It is humbling, and also relieving, to know that we are all made of the same stuff.

ROALD HOFFMANN

Why Buy That Theory?

FROM AMERICAN SCIENTIST

The principle known as Ockham's razor holds that the right explanation for any phenomenon will also be the simplest. The Nobel Prize–winning chemist Roald Hoffmann muses on whether simplicity is all that makes for a successful theory.

The theory of theories goes like this: A theory will be accepted by a scientific community if it explains better (or more of) what is known, fits at its fringes with what is known about other parts of our universe and makes verifiable, preferably risky, predictions.

Sometimes it does go like that. So the theory that made my name (and added to the already recognized greatness of the man with whom I collaborated, *the* synthetic chemist of the 20th century, Robert B. Woodward) did make sense of many disparate and puzzling observations in organic chemistry. And "orbital symmetry control," as our complex of ideas came to be called, made some risky predictions. I remember well the day that Jerry Berson sent us his remarkable experimental results on the stereochemistry of the so-called 1,3-sigmatropic shift. It should proceed in a certain way, he reasoned from our theory—a nonintuitive way. And it did.

But much that goes into the acceptance of theories has little to do with rationalization and prediction. Instead, I will claim, what matters is a heady mix of factors in which psychological attitudes figure prominently.

Simplicity

A SIMPLE EQUATION describing a physical phenomenon (better still, many), the molecule shaped like a Platonic solid with regular geometry, the simple mechanism (A→B, in one step)—these have tremendous *aesthetic* appeal, a direct beeline into our soul. They are beautifully simple, and simply beautiful. Theories of this type are awesome in the original sense of the world—who would deny this of the theory of evolution, the Dirac equation or general relativity?

A little caution might be suggested from pondering the fact that political ads patently cater to our psychobiological predilection for simplicity. Is the world simple? Or do we just want it to be such? In the dreams of some, the beauty and simplicity of equations becomes a criterion for their truth. Simple theories seem to validate that idol of science, Ockham's razor. In preaching the poetic conciseness and generality of orbital explanations, I have succumbed to this, too.

A corrective to the infatuation of scientists with simplicity might come from asking them to think of what they consider beautiful in art, be it music or the visual arts. Is it Bach's *Goldberg Variations* or a dance tune where the theme plays ten times identically in succession? Is any animal ever *painted* to show its bilateral symmetry?

Still, there's no getting away from it; a theory that is simple yet explains a lot is usually accepted in a flash.

Storytelling

WHAT IF the world is complex? Here, symmetry is broken; there, the seemingly simplest of chemical reactions, hydrogen burning to water, has a messy mechanism. The means by which one subunit of hemoglobin communicates its oxygenation to a second subunit, an essential task, resembles a Rube Goldberg cartoon. Not to speak of the intricacies of *any* biological response, from the rise of blood pressure or release of adrenaline when a snake lunges at us, to returning a Ping-Pong serve with backspin. Max Perutz's theory of the cooperativity of oxygen uptake, the way the ribosome functions—these require complicated explanations. And yes, the inherent tinkering of evolution has made them complex. But simpler chemical reactions—a candle burning—are also intricate. As complex as the essential physics of the malleability, brittleness and hardness of metals. Or the geology of hydrothermal vents.

When things are complex yet understandable, human beings weave stories. We do so for several reasons: A→B requires no story. But A→B→C→D and not A→B→C'→D *is* in itself a story. Second, as psychologist Jerome Bruner writes, "For there to be a story, something unforeseen must happen." In science the unforeseen lurks around the next experimental corner. Stories then "domesticate unexpectedness," to use Bruner's phrase.

Storytelling seems to be ingrained in our psyche. I would claim that with our gift of spoken and written language, this is the way we wrest pleasure, psychologically, from a messy world. Scientists are no exception. Part of the story they tell is how they got there—the x-ray films measured over a decade, the blind alleys and false leads of a chemical synthesis. It is never easy, and serendipity substitutes for what in earlier ages would have been called the grace of God. In the end, we overcome. This appeals, and none of it takes away from the ingenuity of the creative act.

In thinking about theories, storytelling has some distinct features. There is always a beginning to a theory—modeling assumptions, perhaps unexpected observations to account for. Then, in a mathematically oriented theory, a kind of development section follows. Something is tried; it leads nowhere, or leaves one dissatisfied. So one essays a variation on what had been a minor theme, and—all of a sudden—it soars. Resolution and coda follow. I think of the surprise that comes from doing a Fourier transform, or of seeing eigenvalues popping out of nothing but an equation and boundary conditions.

Sadly, in the published accounts of theories, much of the narrative of the struggle for understanding is left out, because of self-censorship and the desire to show us as more rational than we were. That's okay; fortunately one can still see the development sections of a theoretical symphony as one examines an ensemble of theories, created by many people, not just one, groping towards understanding.

The other place where narrative is rife is in the hypothesis-forming stage of doing science. This is where the "reach of imagination" of science, as Jacob Bronowski referred to it, is explicit. Soon you will be brought down to earth by experiment, but here the wild man in you can soar, think up any crazy scheme. And, in the way science works, if you are too blinded by your prejudices to see the faults in your theoretical fantasies, you can be sure others will.

Many theories are popular because they tell a rollicking good story, one that is sage in capturing the way the world works, and could be stored away to deal with the next trouble. Stories can be funny; can there be humorous theories?

A Roll-on Suitcase

THEORIES THAT SEEK acceptance had better be *portable*. Oh, people will accept an initiation ritual, a tough-to-follow manual to mastering a theory. But if every application of the theory requires consultation with its originator (that's the goal of commercialization, antithetical to the ethic of science), the theory will soon be abandoned. The most popular theories in fact are those that can be applied by *others* to obtain surprising results. The originator of the theory might have given an eyetooth to have done it earlier, but friends should hold him back—it's better if someone else does it. And cites you.

Relatively uncomplicated models that admit an analytical solution play a special role in the acceptance and popularity of theories among other theorists. I think of the harmonic oscillator, of the Heisenberg and Hückel Hamiltonians, of the Ising Model, my own orbital interactions. The models become modules in a theoretical Erector set, shuttled into any problem as a first (not last) recourse. In part this is fashion, in part testimony to our predilection for simplicity. But, more significantly, the use of soluble models conveys confidence in the value of metaphor—taking one piece of experience over to another. It's also evidence of an existential desire to try something—let's try this.

Productivity

THE BEST THEORIES are productive, in that they stimulate experiment. Science is a wonderfully interactive way for gaining reliable knowledge. What excitement there is in person A advancing a view of how things work, which is tested by B, used by C to motivate making a molecule that tests the limits of the theory, which leads to D (not C) finding that molecule to be superconducting or an antitumor agent, whereupon a horde of graduate students of E or F are put to making slight modifications! People need reasons for doing things. Theories provide them, surely to test the theories (with greater delight if proved wrong), but also just to have a reason for making the next molecule down the line. Theories that provoke experiment are really valued by a community that in every science, even physics, is primarily experimental.

A "corollary" of the significance of productivity is that theories that are fundamentally untenable or ill-defined can still be immensely productive. So was phlogiston in its day, so in chemistry was the idea of resonance energies, calculated in a Hückel model. People made tremendous efforts to make molecules that would never have been made (and found much fascinating chemistry in the process) on the basis of "resonance energies" that had little connection to stability, thermodynamic or kinetic. Did it matter that Columbus miscalculated in his "research proposal" how far the Indies were?

As Jerry Berson has written, "A lot of science consists of permanent experimental facts established in tests of temporary theories."

Frameworks for Understanding

STEPHEN G. BRUSH HAS recently studied a range of fields and discoveries, to see what role predictions play in the acceptance of theories. Here's what he has to say about the new quantum mechanics: "Novel predictions played essentially no role in the acceptance of the most important physical theory of the 20th century, quantum mechanics. Physicists quickly accepted that theory because it provided a coherent deductive account of a large body of known empirical facts . . ." Many theories predict relatively little (quantum mechanics actually did eventually) yet are accepted because they carry tremendous explanatory power. They do so by classification, providing a framework (for the mind) for ordering an immense amount of observation. This is what I think 20th-century theories of acidity and basicity in chemistry (à la Lewis or Brønsted) do. Alternatively, the understanding provided is one of mechanism—this is the strength of the theory of evolution.

It is best to distinguish the concepts of theory, explanation and understanding. Or to try to do so, for they resist differentiation. Evelyn Fox Keller, who in her brilliant book, *Making Sense of Life*, has many instructive tales of theory acceptance, says this of explanation:

> A description or a phenomenon counts as an explanation . . . if and only if it meets the needs of an individual or a community. The challenge, therefore, is to understand the needs that different kinds of explanations meet. Needs do of course vary, and inevitably so: they vary not only with the state of the science at a particular time, with local technological, social, and economic opportunities, but also with larger cultural preoccupations.

As Bas van Fraassen has incisively argued, any explanation is an answer. If we accept that, the nature of the question becomes of essence, and so does our reception of the answer. Both (the reconstructed question of "why?" and our response) are context-dependent and subjective. Understanding, van Fraassen says, "consists in being in a position to explain." And so is equally subjective in a pragmatic universe.

Incidentally, explanations are almost always stories. Indeed, moralistic and deterministic stories. For to be satisfying they don't just say A→B→C→D, but A→B→C→D *because* of such and such propensities of A, B and C. The implicit strong conviction of causality, justified by seemingly irrefutable reason, may be dangerously intoxicating. This is one reason why I wouldn't like scientists and engineers to run this world.

The acceptance of theories depends as much on the psychology of human beings as on the content of the theories. It is human beings who decide, individually and as a community, whether a theory indeed has explanatory power or provides understanding. This is why seemingly "extrascientific" factors such as productivity, portability, storytelling power and aesthetics matter. Sometimes it takes a long time (witness continental drift), but often the acceptance is immediate and intuitive—it fits. Like a nice sweater.

'Tis a Gift

THERE IS something else, even more fundamentally psychological, at work. Every society uses gifts, as altruistic offerings but more importantly as a way of mediating social interactions. In science the gift is both transparent and central. Pure science is as close to a gift economy as we have, as Jeffrey Kovac has argued. Every article in our open literature is a gift to all of us. Every analytical method, every instrument. It's desired that the gift be beautiful (simple gifts are, but also those that bring us a good story with them), to be sure. But that the offering be useful (portable, productive) endows it with special value. The giver will be remembered, every moment, by the one who received the gift.

The purpose of theory, Berson writes, is "to bring order, clarity, and predictability to a small corner of the world." That suffices. A theory is then a special gift, a gift for the mind in a society (of science, not the world) where thought and understanding are preeminent. A gift from one human being to another, to us all.

LEONARD CASSUTO

Big Trouble in the World of "Big Physics"

FROM *SALON*

In September 2002, a committee that was formed to investigate allegations of misconduct against a high-flying young physicist at Bell Labs, Jan Hendrik Schön, concluded that he had fabricated results in several papers published in distinguished journals, invalidating the results of his cutting-edge research on molecular electronics. The committee's report exposed deep flaws throughout the system—from how researchers are recruited by big institutions to the peer review process that governs scientific publication. Leonard Cassuto, who frequently reports on academic politics, investigates how such a scandal could have happened.

In February 2000, a promising young physicist named Jan Hendrik Schön published some startling experimental results. Schön and his partners had started with molecules that don't ordinarily conduct electricity and claimed they had succeeded in making them behave like semiconductors, the circuits that make computers work. The researchers reported their findings in *Science,* one of the flagship scientific journals.

The data created an immediate stir. Schön, who works at Lucent Technologies' prestigious Bell Labs, followed that paper up with another, and then another. In his world of "publish or perish," he became a virtual writing machine, issuing one article after another. His group reported that they could make other nonconductors into semiconductors, lasers and light-absorbing devices. These claims were revolutionary. Their implications for electronics and other fields were enormous, holding the promise that computing circuitry might one day shrink to unimaginably small size. In the words of one Princeton professor, Schön had "defeated chemistry." He had become a modern alchemist, apparently conducting electricity where it had never gone before.

In a field where publishing two or three articles a year makes you productive, Schön started issuing reports in bunches. He was the lead author on dozens of articles—more than 90 in about three years, most of them appearing in the industry-leading journals. In 2001, he received an award for scientific "Breakthrough of the Year," but most scientists saw this recognition as only the beginning.

"I saw these results being presented to a German audience," says James Heath of UCLA, "and they knock on the chairs instead of clapping. It was incredible—they got a 'standing knocking.' I thought, These guys are going to Stockholm." Less than five years after finishing graduate school, Jan Hendrik Schön was in contention for the Nobel Prize.

Then the wunderkind fell to earth. In April 2002, a small group of researchers at Bell Labs contacted Princeton physics professor Lydia Sohn and whispered that all was not right with Schön's data. Sohn recalls that she and Cornell University's Paul McEuen stayed up late one night and found some disturbing coincidences in Schön's results: The same graphs were being used to illustrate the outcomes of completely different experiments. "You would expect differences," she said, "but the figures were identical. It was a smoking gun."

Once tipped off, McEuen started looking closely at a range of Schön's work, enlarging the graphs and playing a game of mix-and-match. He found many duplicate graphs in different papers on different subjects. Schön was apparently using the same sets of pictures to tell lots of different stories.

In May, McEuen and Sohn formally alerted the editors of *Science* and *Nature*—where Schön and his team had published numerous articles—of the discrepancies. McEuen and Sohn also informed Schön; his supervisor and coauthor, Bertram Batlogg; and Bell Labs management that they were blowing the whistle. Schön immediately insisted that his experiments were fine, and that the duplicated figures were a simple clerical error for which he now offered substitutes. To *Nature* he declared he was "confident" of his results. To *Science* he said, "I haven't done anything wrong." Batlogg mostly said nothing at all. A scandal had broken out in the world of physics.

Lucent Technologies, which runs Bell Labs, responded swiftly. Cherry Murray, head of physical science research, acted with other Bell Labs officials and appointed an independent committee to look into the matter. The panel was made up primarily of university physics professors, led by Malcolm Beasley of Stanford. Their mandate, according to Beasley, is to get the facts and "find out whether scientific misconduct has occurred."

"Big Physics" is a small world. Very few people can understand, let alone judge, what experimental physicists do. They work in close professional communities of specialists and subspecialists, conducting expensive experiments and publishing papers with names like "Gate-Induced Superconductivity in a Solution-Processed Organic Polymer Film."

But physics is also a field in which millions of taxpayer dollars are spent every year. Now physics has an accountability problem and the only possible auditors are other physicists. As the field reels from what may be the biggest fraud in its history, scientists across the world are alarmed: Bad science can cost lives—think of the untested O-rings on the space shuttle *Challenger* that froze stiff and caused the ship's tragic explosion. But what about phony science?

Jan Hendrik Schön joined Bell Labs in 1998, just before finishing his Ph.D. in Konstanz, Germany. His international move was typical; the physics community is a far-flung network within which virtually all practicing researchers have connections to specialists in other countries.

But if physics is global, the United States is its financial center. There are more scientists doing expensive experiments in the U.S. than in any other country. Most work at universities as professors, but walking in step with faculty members, attending the same conferences and publishing in the same journals are corporate-funded researchers at places like Xerox, IBM and Bell Labs.

Like university departments, science labs operated by giant corporations depend on income from the larger entity (the university maintains its departments, while the corporation maintains its lab). Both also receive government money, often to conduct joint ventures. Together, the schools and the corporations make up one large academic community.

Bell Labs, formerly operated by AT&T, is the most famous of all corporate science centers. In 77 years of existence, the Labs have hired top-flight scientists from universities and essentially turned them loose to look into whatever they've wanted, with the corporation footing the bill. If their discoveries had practical use, that was great. Otherwise, the science was, like much university-based research, a contribution to common knowledge.

Researchers at Bell Labs were like professors without teaching and other administrative responsibilities. Given up-to-date equipment, funding and generous salaries, these scientists were pointed in the direction of the unknown and encouraged to work together to explore it.

The results of this policy have been impressive: Bell Labs scientists have won numerous Nobel Prizes and other awards. But since AT&T decided in 1996 to split into software and hardware companies—with the latter, Lucent Technologies, retaining Bell Labs—the facility has fallen upon hard times. The Schön affair is a black eye on an already battered company. In one quarter, Lucent lost a staggering $8 billion and laid off thousands of employees.

Schön himself was set to leave Bell Labs, to become a director at the Max Planck Institute in Germany, but the job offer was withdrawn when the scandal broke. When the news of the duplicated graphs first became public, Schön defended himself vigorously until Lucent imposed a gag order on all its employees about the matter. After that, he could only sit in silent limbo, waiting for the Beasley panel to issue its findings.

The duplicated graphs are not the only smoking gun. There's also the serious problem that despite numerous attempts, no other physicist has repeated Schön's results. If no one else can repeat the results of an experiment, both experiment and experimenter come under suspicion. "It is part of the process of science," says investigative committee head Beasley, "that things get winnowed out because they don't work."

Physicist Art Ramirez of Los Alamos National Laboratory once told *Science* that Schön had "magic hands." Now, says Ramirez, "I'm less sure. I'm getting less comfortable" with Schön's work. Schön himself appears to have lost his magic touch. He told *Science* in the wake of the controversy that he was "trying as hard as [I] can" to duplicate his own results, but somehow the experiments don't work for him anymore.

They haven't been working for other scientists, either. Physicists around the country and the world have spent tens of millions of dollars—including funding from the U.S. Department of Energy—trying to reproduce Schön's key results. Taxpayers have footed the bill for two years' worth of fruitless and expensive efforts. "It seemed so plausible," sighs Arthur Hebard of the University of Florida. "Almost too good to be true." Now Hebard wonders, "What's the trick?"

There are an estimated 100 laboratory groups working on Schön's results in the United States and around the world. For graduate students basing their Ph.D. research on Schön's experiments, their education is at stake. Postdoctoral fellows worry about their prospects for future employment. Some junior professors have tied their bids for tenure to experiments based on Schön's findings. Their professional livelihoods are literally at risk. If the results are fake, how can these people get their careers back? Invoking recent headlines, UCLA's Heath commented that "this is like the opposite of losing your retirement." Asked one nervous faculty member, "Can we get a class action suit together?"

When Martin Fleischmann and B. Stanley Pons suddenly walked out of the University of Utah chemistry department in 1989 claiming that they had solved our energy problems by producing a "cold fusion" reaction (the heat of such reactions has reserved them for hydrogen bombs), scientists showed by straightforward calculation that the experiment couldn't work. Not surprisingly, no one could repeat the results the two claimed. Though the matter received a lot of media coverage, it was a case of routine exposure of a couple of unknowns.

Schön's work has also never crossed the repeatability threshold. Skepticism about it was rising before the scandal broke. By the time his colleague McEuen helped find the duplications, says Cornell's Dan Ralph, "We were having serious doubts about the science." UCLA's Heath described how when a Schön paper would come out, he would get excited, but after a while "I would begin worrying a little bit." Sohn, who worked with McEuen to make the matter public, says, "The data were too clean. They were what you'd expect theoretically, not experimentally. People were getting frustrated because no one could reproduce the results, and it was hitting a crescendo."

Many physicists now wonder about Schön's incredible productivity. "I am guilty of extreme gullibility," says Nobel laureate Philip Anderson. "I have to confess it. We should all have been suspicious of the data almost immediately." Ramirez of Los Alamos says, "I find it hard to even read that many papers, much less write them."

Why would Schön rush to publish dubious results if he knew others would attempt to repeat his experiments? Perhaps, says Heath, Schön was "innocent and naive," like Utah's Fleischmann and Pons. One physicist gave voice to a darker possibility: "If the results are fraudulent, Schön would have to have some kind of psychological problem."

Like other academic fields, physics polices itself through a peer review system. When a physicist submits a paper for publication, the editor sends it out to be judged by specialists in the author's field. These referees recommend publication (sometimes with revision) or rejection. The system is designed to weed out substandard work, and to improve promising submissions and make them publishable. It's supposed to keep things honest.

Peer review also governs external funding. Experimental physicists need labs to work in, and the equipment in a typical condensed-matter physics lab costs about a million dollars. Further funds are required for upkeep, and scientists and their staff need salaries. Universities maintain a lot of the country's physics labs and pay much of the cost out of tuition and endowment income, but an important part of any physics professor's job is to look for additional funding. Corporations are one source, and in cases like Bell Labs, the parent corporation pays most of the researchers' bills.

Perhaps the biggest single source of funding for scientific research is the taxpayer. The federal government dispenses about $20 billion a year to scientists and mathematicians through numerous outlets. The National Science Foundation is the most abundant source, awarding about $5 billion annually. The Department of Defense also supports many a physics lab, as do NASA and the Department of Energy. How does the government decide who gets the money? It invites physicists to Washington to read their colleagues' grant applications and make the judgments. "There's a certain amount of trust in the physicists," said Jonathan Epstein, science advisor to New Mexico Sen. Jeff Bingaman, chairman of the Senate Science and Energy Committee. The peer review system is the means by which that trust is maintained.

The Schön affair has besmirched the peer review process in physics as never before. Why didn't the peer review system catch the discrepancies in his work? A referee in a new field doesn't want to "be the bad guy on the block," says Dutch physicist Teun Klapwijk, so he generally gives the author the benefit of the doubt. But physicists did become irritated after a while, says Klapwijk, "that Schön's flurry of papers continued without increased detail, and with the same sloppiness and inconsistencies."

Some critics hold the journals responsible. The editors of *Science* and *Nature* have stoutly defended their review process in interviews with the London *Times Higher Education Supplement.* Karl Ziemelis, one of *Nature*'s physical science editors, complained of scapegoating, while Donald Kennedy, who edits *Science,* asserted that "there is little journals can do about detecting scientific misconduct."

Maybe not, responds Nobel Prize–winning physicist Philip Anderson of Princeton, but the way that *Science* and *Nature* compete for cutting-edge work "compromised the review process in this instance." These two industry-leading publications "decide for themselves what is good science—or good-selling science," says Anderson (who is also a former Bell Labs director), and their market consciousness "encourages people to push into print with shoddy results." Such urgency would presumably lead to hasty review practices. Klapwijk, a superconductivity specialist, said that he had raised objections to a Schön paper sent to him for review, but that it was published anyway.

Klapwijk points out that the duplicated figures were in separate papers that weren't necessarily sent to the same people for vetting. But as one physicist admits, "It's hard to criticize someone else's productivity without sounding like you're full of sour grapes."

Another reason for the breakdown is the hypnotizing effect of reputation. When the names of eminent people and places appear on the top of submitted papers, says Florida physicist Hebard, "reviewers react almost unconsciously" to their prestige. "People discount reports from groups that aren't well known," adds University of Maryland physicist Richard Greene.

"Part of the reason the work was accepted," says Greene, was because Schön's coauthor and one-time supervisor Bertram Batlogg put his imprimatur (and that of Bell Labs) on it. Batlogg has been a respected superconductivity physicist for more than two decades.

Batlogg left Bell Labs for a job in Switzerland before he became a cause célèbre. He now stands accused of harboring, if not abetting, scientific fraud. In his only public pronouncement about the scandal, in a German magazine, Batlogg said, "If I'm a passenger in a car that drives through a red light, then it's not my fault."

Most other scientists feel very differently. "People don't want to hear this. They want to hear a mea culpa. Batlogg allowed this to happen," says Art Ramirez of Los Alamos. "Batlogg signed on," Hebard says. "He's a collaborator, not a casual passenger. He's been benefiting all along, riding the public wave." Adds Princeton's Sohn, "If a young driver has a learner's permit, then who's responsible for him? Batlogg was the licensed driver, and Schön was the student driver."

"If my student came to me with earth-shattering data, you wouldn't be able to pry me out of the lab," says Rice University's Douglas Natelson. "I'd be in there turning the knobs myself." Heath echoes this sentiment: "I'd sit down there to see how this is being done. I'd demand to see it several times."

Siegfried Grossman, head of a German research consortium, told a German publication that Batlogg is simply making excuses. Coauthors, Grossman said, must take full responsibility for the contents of their publications. Sohn says flatly, "I am responsible for what my students publish. If my name is going to be on a paper, I want to make sure it's right."

Batlogg recruited Schön while Schön was still a graduate student. He brought Schön into his lab. He sponsored Schön's experiments. And rather than formally withdraw any papers he might have considered suspicious, he gave many well-received talks at elite international conferences on the results. Wonders one American physicist, "What did Batlogg know and when did he know it? I don't see how he can work as a scientist any longer." Added Allen Goldman of the University of Minnesota, "Batlogg's going to take his lumps on this one."

What do we as a society expect from our scientists? We equate the scientific method with abstract inquiry, but as biologist Stephen Jay Gould was fond of pointing out, you have to be looking for something in the first place—and your goal is bound to affect your search. Science, Gould suggested, involves a balancing act between objective methods and subjective goals.

There is one shining rule, though: no cheating. Science, like any academic field, demands scrupulous, rational honesty. "My goal may be to win a prize," says Nobel laureate Horst Stormer, "but my duty is to report what I have observed in the most objective way that I can. I say this in the strongest terms. This is what I expect from my colleagues, from my graduate students, at all levels of the field."

American intellectual culture hasn't exactly been showcasing that sort of rectitude and responsibility lately. The late Stephen Ambrose and Doris Kearns Goodwin, two historians who admitted to plagiarism in their books, have seen their individual reputations suffer for their acts, and they've tainted their discipline at the same time. Now we may have to make room for another in the public stocks. Schön, his colleagues say, is also risking the reputation of an entire field.

Physicists everywhere are relying heavily on the Beasley committee to set things right. Some hope to polish tarnished reputations. Christian Kloc, for example, is a chemist on the Schön team whose job was to supply tiny crystals for the experiments. Kloc's work appears to be unrelated to the disputed data, but as one physicist put it, "Who knows anymore?" But there is more at stake than the careers of individuals. If the accusations turn out to be true, says Cornell's Dan Ralph, "This is the biggest fraud in the history of modern physics."

McEuen, the man who helped to expose the problem, has confidence in the investigation. Beasley himself is more circumspect. Acknowledging that the physics community may be expecting more from his committee's report than its mandate suggests, Beasley says only that "at the end of the day, we need to demonstrate that we took this very seriously and that we did a good job."

More immediately, Dan Ralph of Cornell remains concerned about the careers of younger physicists that may have been jeopardized, and by the unreliability the whole system now shows. "Checks and balances didn't work the way they should have," he said. As a result, "The fallout from this will hurt," according to Hebard. Many fear that Bell Labs will not recover. Because Schön's results are now suspect, Hebard and other scientists worry that funding for a highly promising area will now dry up. But Hebard sees the effect of the scandal extending beyond the matter of organic superconductivity. "We thought we were inviolate," Hebard said. "Scientists are easy to fool because you believe what your colleagues tell you. I would hope that the public wouldn't conflate this with Enron and WorldCom, but it is inflating the profit statement."

And when the news reaches the nation's high school physics classrooms? "Science is scientists," said William Wallace, teacher and head of the science department at Washington's Georgetown Day School. "It's a human activity." Still, Wallace concedes that "A little trust is chipped away every time something like this happens." Pointing to the "heroes I had growing up"—like Richard Feynman, the maverick Nobel Prize winner who inspired generations of physics students—Wallace notes that now "there's an incredible amount of pressure on young and midcareer scientists. They always need to know where the next grant is coming from." The result is "careerism," not heroism or pursuit of the truth. And that leaves the teacher with a question: "In the end, if there isn't respect for scientific truth, then what have you got?"

Hawking's Breakthrough Is Still an Enigma

FROM *THE NEW YORK TIMES*

Stephen Hawking, perhaps the world's most famous scientist, is known to the world for his best-selling books and for his brilliant mind, undimmed by his failing body. In the world of physics he is known for something else—his startling discoveries about the nature of those most mysterious celestial objects, black holes. Thirty years after his key insight, scientists are still grappling with its implications. Dennis Overbye reports on their progress.

In the fall of 1973 Dr. Stephen W. Hawking, who has spent his entire professional career at the University of Cambridge, found himself ensnared in a horrendous and embarrassing calculation. Attempting to investigate the microscopic properties of black holes, the gravitational traps from which not even light can escape, Dr. Hawking discovered to his disbelief that they could leak energy and particles into space, and even explode in a fountain of high-energy sparks.

Dr. Hawking first held off publishing his results, fearing he was mistaken. When he reported them the next year in the journal *Nature*, he titled his paper simply "Black Hole Explosions?" His colleagues were dazzled and mystified. Nearly 30 years later, they are still mystified. When they gathered in Cambridge in January 2002, to mark Dr. Hawking's 60th birthday with a weeklong workshop titled "The Future of Theoretical Physics and Cosmology," the ideas spawned by his calculation and its aftermath often took center stage.

They are ideas that touch on just about every bone-jarring abstruse concept in modern physics.

"Black holes are still fundamentally enigmatic objects," said Dr. Andrew Strominger, a Harvard physicist, who attended. "In fundamental physics, gravity and quantum mechanics are the big things we don't understand. Hawking's discovery of black hole radiation was of fundamental importance to that connection."

Black holes are the prima donnas of Einstein's general theory of relativity, which explains the force known as gravity as a warp in space-time caused by matter and energy. But even Einstein could not accept the idea that the warping could get so extreme, say in the case of a collapsing star, that space could wrap itself completely around some object like a magician's cloak, causing it to disappear as a black hole.

Dr. Hawking's celebrated breakthrough resulted partly from a fight. He was hoping to disprove the contention of Jacob Bekenstein, then a graduate student at Princeton and now a professor at the Hebrew University in Jerusalem, that the area of a black hole's boundary, the point of no return in space, was a measure of the entropy of a black hole. In thermodynamics, the study of heat and gases, entropy is a measure of wasted energy or disorder, which might seem like a funny concept to crop up in black holes. But in physics and computer science, entropy is also a measure of the information capacity of a system—the number of bits that it would take to describe its internal state. In effect, a black hole or any other system was like a box of Scrabble letters—the more letters in the box the more words you could make, and the more chances of gibberish.

According to the second law of thermodynamics, the entropy of a closed system always stays the same or increases, and Dr. Hawking's own work had shown that the hole's surface area always increased, a process that seemed to ape that law.

But Dr. Hawking, citing classical physics, argued that an object with entropy had to have a temperature, and anything with a temperature—from a fevered brow to a star—must radiate heat and light with a characteristic spectrum. If a black hole could not radiate, it could have no temperature and thus no entropy. But that was before gravity, which shapes the cosmos, met quantum theory, the paradoxical rules that describe the behavior of matter and forces within it. When Dr. Hawking added a touch of quantum uncertainty to the standard Einsteinian black hole model, particles started emerging. At first he was annoyed, but when he realized this "Hawking radiation" would have the thermal spectrum predicted by thermodynamic theory, he concluded his calculation was right.

But there was a problem. The radiation was random, Dr. Hawking's theory said. As a result, all the details about whatever had fallen into the black hole could be completely erased—a violation of a hallowed tenet of quantum theory, which holds that it should always be possible to run the film backwards and find out the details of how something started—whether an elephant or a Volkswagen had been tossed into the black hole, for example. If he was right, Dr. Hawking suggested, quantum theory might have to be modified. Black holes, he said in his papers and talks in the late 1970s, were ravagers of information, spewing indeterminacy and undermining law and order in the universe.

"God not only plays dice with the universe," Dr. Hawking said, inverting the phrase by which Einstein had famously rejected quantum uncertainty, "but sometimes throws them where we can't see them." Such statements aroused the attention of particle physicists. Weird as it may be, quantum theory is nonetheless the foundation on which much of the modern world is built, everything from transistors to CDs, and it is the language in which all of the fundamental laws of physics, save gravity, are expressed. "This cannot be," Dr. Leonard Susskind, a theorist at Stanford, recalled saying to himself.

It was the beginning of what Dr. Susskind calls an adversarial relationship. "Stephen Hawking is one of the most obstinate people in the world; no, he is the most infuriating person in the universe," Dr. Susskind told the birthday workshop, as Dr. Hawking grinned in the back row.

In the ensuing 20 years, opinions have split mostly along party lines. Particle physicists like Dr. Susskind and Dr. Gerard 't Hooft, a physicist at the University of Utrecht and the 1999 Nobel Prize winner, defend quantum theory and say that the information must get out somehow, perhaps subtly encoded in the radiation. Another possibility—that the information was left behind in some new kind of elementary particle when the black hole evaporated—seems to have fallen from favor.

Relativity experts like Dr. Hawking and his friend the Caltech physicist Dr. Kip Thorne were more likely to believe in the power of black holes to keep secrets. In 1997, Dr. Hawking and Dr. Thorne put their money where the black hole mouth was, betting Dr. John Preskill, a Caltech particle physicist, a set of encyclopedias that information was destroyed in a black hole.

To date neither side has felt obliged to pay up.

Writing on the Wall

DR. SUSSKIND AND OTHERS have argued that nothing ever makes it into the black hole to begin with because, in accord with Einstein, everything at the boundary, where time slows, would appear to an outside observer to "freeze" and then fade, spreading out on the surface where it could produce subtle distortions in the Hawking radiation.

In principle, then, information about what had fallen onto the black hole could be read in the radiation and reconstructed; it would not have disappeared.

The confusion had arisen, Dr. Susskind explained, because physicists had been trying to imagine the situation from the viewpoint of God rather than that of a particular observer who had to be either in the black hole or outside, but not both places at once. When the accounting is done properly, he said, "No observer sees a violation of the laws of physics."

The information paradox made it important for theorists to try to go beyond thermodynamic analogies and actually calculate how black holes store information or entropy. But there was a catch. According to a well-known formula developed by the Austrian physicist Ludwig Boltzmann (and engraved on his tombstone), the entropy of a system could be determined by counting the number of ways its contents could be arranged.

In order to enumerate the possible ways of arranging the contents of a black hole, physicists needed a theory of what was inside. By the mid-1990s they had one: string theory, which portrays the forces and particles of nature, including those responsible for gravity, as tiny vibrating strings.

In this theory, a black hole is a tangled melange of strings and multidimensional membranes known as "D-branes." In a virtuoso calculation in 1995, Dr. Strominger and Dr. Cumrun Vafa, also of Harvard, untangled the innards of an "extremal" black hole, in which electrical charge just balanced gravity.

Such a hole would stop evaporating and would thus appear static, allowing the researchers to count its quantum states. They calculated that the entropy of a black hole was its area divided by four—just as Dr. Hawking and Dr. Bekenstein said it would be.

The result was a huge triumph for string theory. "If string theory had been wrong, that would have been deadly," Dr. Strominger said.

The success of the Harvard calculation has encouraged some particle physicists to conclude that black holes can be analyzed with the tools of quantum mechanics, and thus that the information issue has been resolved. But others say this has yet to be accomplished—among them Dr. Strominger, who added, "It remains an unsettled issue."

Degrees of Freedom

PERHAPS THE MOST MYSTERIOUS and far-reaching consequence of the exploding black hole is the idea that the universe can be compared to a hologram, in which information for a three-dimensional image can be stored on a flat surface, like an image on a bank card.

In the 1980s, extending his and Dr. Hawking's work, Dr. Bekenstein showed that the entropy and thus the information needed to describe any object were limited by its area. "Entropy is a measure of how much information you can pack into an object," he explained. "The limit on entropy is a limit on information."

This was a strange result. Normally you might think that there were as many choices—or degrees of freedom about the inner state of an object—as there were points inside that space. But according to the so-called Bekenstein bound, there were only as many choices as there were points on its outer surface.

The "points" in this case are regions with the dimensions of 10–33 centimeters, the so-called Planck length, that physicists believe are the "grains" of space. According to the theory, each of these can be assigned a value of zero or one—yes or no—like the bits in a computer.

"What happens when you squeeze too much information into an object is that you pack more and more energy in," said Rafael Bousso, a physicist at the Kavli Institute for Theoretical Physics at the University of California, Santa Barbara. But if it gets too heavy for its size, it becomes a black hole, and then "the game is over," as he put it. "Like a piano with lots of keys but you can't press more than five of them at once or the piano will collapse."

The holographic principle, first suggested by Dr. 't Hooft in 1993 and elaborated by Dr. Susskind a year later, says in effect that if you can't use the other piano keys, they aren't really there. "We had a completely wrong picture of the piano," explained Dr. Bousso. The normal theories that physics uses to describe events in space-time are redundant in some surprising and as yet mysterious way. "We clearly see the world the way we see a hologram," Dr. Bousso said. "We see three dimensions. When you look at one of those chips, it looks pretty real, but in our case the illusion is perfect."

Dr. Susskind added: "We don't read the hologram. We are the hologram."

The holographic principle, these physicists say, can be applied to any space-time, but they have no idea why it works.

"It really should be mysterious," Dr. Strominger said. "If it's really true, it's a deep and beautiful property of our universe—but not an obvious one."

The Frontiers of Beauty

THAT BEAUTY, however, comes at a price, said Dr. 't Hooft, namely cause and effect. If the information about what we are doing resides on distant imaginary walls, "how does it appear to us sitting here that we are obeying the local laws of physics?" he asked the audience at the Hawking birthday workshop.

Quantum mechanics had been saved, he declared, but it still might need to be supplanted by laws that would preserve what physicists call "naive locality."

Dr. 't Hooft acknowledged that there had been many futile attempts to eliminate quantum mechanics' seemingly nonsensical notions, like particles that can instantaneously react to one another across light-years of space. In each case, however, he said there were assumptions, or "fine print," that might not hold up in the end.

Recent observations have raised the stakes for ideas like holography and black hole information. The results suggest that the expansion of the universe is accelerating. If it goes on, astronomers say, distant galaxies will eventually be moving away so fast that we will not be able to see them anymore.

Living in such a universe is like being surrounded by a horizon, glowing just like a black hole horizon, over which information is forever disappearing. And since this horizon has a finite size, physicists say, there is a limit to the amount of complexity and information the universe can hold, ultimately dooming life.

Physicists admit that they do not know how to practice physics or string theory in such a space, called a de Sitter space after the Dutch astronomer Willem de Sitter, who first solved Einstein's equations to find such a space. "De Sitter space is a new frontier," said Dr. Strominger, who hopes that the techniques and attention that were devoted to black holes in the last decade will enable physicists to make headway in understanding a universe that may actually represent the human condition.

Dr. Bousso noted that it was only in the last few years, with the discovery of D-branes, that it had been possible to solve black holes. What other surprises await in string theory? "We have no idea how small or large a piece of the theory we haven't seen yet," he said.

In the meantime, perhaps in imitation of Boltzmann, Dr. Hawking declared at the end of the meeting that he wanted the formula for black hole entropy engraved on his own tombstone.

RICHARD C. LEWONTIN
AND RICHARD LEVINS

Stephen Jay Gould: What Does It Mean to Be a Radical?

FROM *MONTHLY REVIEW*

A great scientist who also strived to reach the wider public, Stephen Jay Gould achieved the nigh-impossible, doing groundbreaking work in evolutionary biology while writing essays and books of great elegance and broad popular appeal. In this appreciation of his life and career, two of his colleagues at Harvard reflect on how Gould's work demonstrated the value of taking a radical approach to science.

In early 2002, Stephen Gould developed lung cancer, which spread so quickly that there was no hope of survival. He died on May 20, 2002, at the age of sixty. Twenty years earlier, he had escaped death from mesothelioma, induced, we all supposed, by some exposure to asbestos. Although his cure was complete, he never lost the consciousness of his mortality and gave the impression, at least to his friends, of an almost cheerful acceptance of the inevitable. Having survived one cancer that was probably the consequence of an environmental poison, he succumbed to another.

The public intellectual and political life of Steve Gould was extraordinary, if not unique. First, he was an evolutionary biologist and historian of science whose intellectual work had a major impact on our views of the process of evolution. Second, he was, by far, the most widely known and influential expositor of science who has ever written for a lay public. Third, he was a consistent political activist in support of socialism and in opposition to all forms of colonialism and oppression. The figure he most closely resembled in these respects was the British biologist of the 1930s, J. B. S. Haldane, a founder of the modern genetical theory of evolution, a wonderful essayist on science for the general public, and an idiosyncratic Marxist and columnist for *The Daily Worker* who finally split with the Communist Party over its demand that scientific claims follow Party doctrine.

What characterizes Steve Gould's work is its consistent radicalism. The word *radical* has come to be synonymous with *extreme* in everyday usage: *Monthly Review* is a *radical* journal to the readers of *The Progressive*; Steve Gould underwent *radical* surgery when tumors were removed from his brain; and a *radical* is someone who is out in left (or right) field. But a brief excursion into the *Oxford English Dictionary* reminds us that the root of the word radical is, in fact, *radix*, the Latin word for *root*. To be radical is to consider things from their very root, to go back to square one, to try to reconstitute one's actions and ideas by building them from first principles. The impulse to be radical is the impulse to ask, "How do I know that?" and, "Why am I following this course rather than another?" Steve Gould had that radical impulse and he followed it where it counted.

First, Steve was a radical in his science. His best-known contribution to evolutionary biology was the theory of *punctuated equilibrium* that he developed with his colleague Niles Eldridge. The standard theory of the change in the shape of organisms over evolutionary time is that it occurs constantly, slowly, and gradually with more or less equal changes happening in equal time intervals. This seems to be the view that Darwin had, although almost anything can be read from Darwin's nineteenth-century prose. Modern genetics has shown that any heritable change in development that is at all likely to survive will cause only a slight change in the organism, that such mutations occur at a fairly constant rate over long time periods and that the force of natural selection for such small changes is also of small magnitude. These facts all point to a more or less constant and slow change in species over long periods.

When one looks at the fossil record, however, observed changes are much more irregular. There are more or less abrupt changes in shape between fossils that succeed each other in geological time with not much evidence for the supposed gradual intermediates between them. The usual explanation is that fossils are relatively rare and we are only seeing occasional snapshots of the actual progression of organisms. This is a perfectly coherent theory, but Eldridge and

Gould went back to square one, and questioned whether the rate of change under natural selection was really as constant as everybody assumed. By examining a few fossil series in which there was a much more complete temporal record than is usual, they found evidence of long periods of virtually no change punctuated by short periods during which most of the change in shape appeared to occur. They generalized this finding into a theory that evolution occurs in fits and starts and provided several possible explanations, including that much of evolution occurred after sudden major changes in environment. Steve Gould went even further in his emphasis on the importance of major irregular events in the history of life. He placed great importance on sudden mass extinction of species after collisions of large comets with the Earth and the subsequent repopulation of the living world from a restricted pool of surviving species. The temptation to see some simple connection between Steve's theory of episodic evolution and his adherence to Marx's theory of historical stages should be resisted. The connection is much deeper. It lies in his radicalism.

Another aspect of Gould's radicalism in science was in the form of his general approach to evolutionary explanation. Most biologists concerned with the history of life and its present geographical and ecological distribution assume that natural selection is the cause of all features of living and extinct organisms and that the task of the biologist, insofar as it is to provide explanations, is to come up with a reasonable story of why any particular feature of a species was favored by natural selection. If, when the human species lost most of its body hair in evolving from its ape-like ancestor, it still held on to eyebrows, then eyebrows must be good things. A great emphasis of Steve's scientific writing was to reject this simplistic Panglossian adaptationism, and to go back to the variety of fundamental biological processes in the search for the causes of evolutionary change. He argued that evolution was a result of random as well as selective forces and that characteristics may be the physical byproducts of selection for other traits. He also argued strongly for the historical contingency of evolutionary change. Something may be selected for some reason at one time and then for an entirely different reason at another time, so that the end product is the result of the whole history of an evolutionary line, and cannot be accounted for by its present adaptive significance. Thus, for instance, humans are the way we are because land vertebrates reduced many fin patterns to four limbs, mammals' hearts happen to lean to the left while birds' hearts lean to the right, the bones of the inner ear were part of the jaw of our reptilian ancestors, and it just happened to get dry in east Africa at a crucial time in our evolutionary history. Therefore, if intelligent life should ever visit us from elsewhere in the universe, we should not expect them to have a human shape, suffer from sexist hierarchy, or have a command deck on their spaceship.

Gould also emphasized the importance of developmental relations between different parts of an organism. A famous case was his study of the Irish elk, a very large extinct deer with enormous antlers, much greater in proportion to the animal's size than is seen in modern deer. The invented adaptationist story was that male deer antlers are under constant natural selection to increase in size because males use them in combat when they compete for access to females. The Irish elk pushed the evolution of this form of machismo too far and their antlers became so unwieldy that they could not carry on the normal business of life and so became extinct. What Steve showed was that for deer in general, species with larger body size have antlers that are more than proportionately larger, a consequence of a differential growth rate of body size and antler size during development. In fact, Irish elk had antlers of exactly the size one would predict from their body size and no special story of natural selection is required.

None of Gould's arguments about the complexity of evolution overthrows Darwin. There are no new paradigms, but perfectly respectable "normal science" that adds richness to Darwin's original scheme. They typify his radical rule for explanation: always go back to basic biological processes and see where that takes you.

Steve Gould's greatest fame was not as a biologist but as an explicator of science for a lay public, in lectures, essays, and books. The relation between scientific knowledge and social action is a problematic one. Scientific knowledge is an esoteric knowledge, possessed and understood by a small elite, yet the use and control of that knowledge by private and public powers is of great social consequence to all. How is there to be even a semblance of a democratic state when vital knowledge is in the hands of a self-interested few? The glib answer offered is that there are instruments of the popularization of science, chiefly science journalism and the popular writings of scientists, which create an informed public. But that popularization is itself usually an instrument of obfuscation and the pressing of elite agendas.

Science journalists suffer from a double disability: First, no matter how well-educated, intelligent, and well-motivated, they must, in the end, trust what scientists tell them. Even a biologist must trust what a physicist says about quantum mechanics. A large fraction of science reporting begins with a press conference or release produced by a scientific institution. "Scientists at the Blackleg Institute announced today the discovery of the gene for susceptibility to repetitive motion injury." Second, the media for which science reporters work put immense pressure on them to write dramatic accounts. Where is the editor who will allot precious column inches to an article about science whose message is that it is all very complicated, that no predictions can be made, that there are serious experimental difficulties in the way of finding the truth of the matter, and that we may never know the answer? Third, the esoteric nature of scientific knowledge places almost insuperable rhetorical barriers between even the most knowledgeable journalist and the reader. It is not generally realized that a transparent explanation in terms accessible to the lay reader requires the deepest possible knowledge of the matter on the part of the writer.

Scientists, and their biographers, who write books for a lay public are usually concerned to press uncritically the romance of the intellectual life, the wonders of their science, and to propagandize for yet greater support of their work. Where is the heart so hardened that it cannot be captivated by Stephen Hawking and his intellectual enterprise? Even when the intention is simply to inform a lay public about a body of scientific knowledge, the complications of the actual state of understanding are so great that the pressure to tell a simple and appealing story is irresistible.

Steve Gould was an exception. His three hundred essays on scientific questions, published in his monthly column in *Natural History Magazine,* many of which were widely distributed in book form, combined a truthful and subtle explication of scientific findings and problems, with a technique of exposition that neither condescended to his readers nor oversimplified the science. He told the complex truth in a way that his lay readers could understand, while enlivening his prose with references to baseball, choral music, and church architecture. Of course, when we consider writing for a popular audience, we have to be clear about what we mean by *popular.* The Uruguayan writer Eduardo Galeano asked what we mean by writing for "the people" when most of our people are illiterate. In the North there is less formal illiteracy, but Gould wrote for a highly educated, even if nonspecialist, audience for whom choral music and church architecture provided more meaningful metaphors than the scientific ideas themselves.

Most of the subjects Steve dealt with were meant to be illustrative precisely of the complexity and diversity of the processes and products of evolution. Despite the immense diversity of matters on which he wrote there was, underneath, a unifying theme: that the complexity of the living world cannot be treated as a manifestation of some grand general principle, but that each case must be understood by examining it from the ground up and as the realization of one out of many material paths of causation.

In his political life Steve was part of the general movement of the left. He was active in the anti–Vietnam War movement, in the work of Science for the People, and of the New York Marxist School. He identified himself as a Marxist but, as with Darwinism, it is never quite certain what that identification implies. Despite our close comradeship in many things over many years, we never had a discussion of Marx's theory of history or of political economy. More to the point, however, by insisting on his adherence to a Marxist viewpoint, he took the opportunity offered to him by his immense fame and legitimacy as a public intellectual to make a broad public think again about the validity of a Marxist analysis.

At the level of actual political struggles, his most important activities were in the fight against creationism and in the campaign to destroy the legitimacy of biological determinism including sociobiology and racism. He argued before the Arkansas State Legislature that differences among evolutionists or unsolved evolutionary problems do not undermine the demonstration of evolution as an organizing principle for understanding life. He was one of the authors of the original manifesto challenging the claim of sociobiology that there is an evolutionarily derived and hard-wired human nature that guarantees the perpetuation of war, racism, the inequality of the sexes, and entrepreneurial capitalism. He continued throughout his career to attack this ideology and show the shallowness of its supposed roots in genetics and evolution. His most significant contribution to the delegitimation of biological determinism, however, was his widely read exposure of the racism and dishonesty of prominent scientists, *The Mismeasure of Man*. Here again, Gould showed the value of going back to square one.

Not content simply to show the evident class prejudice and racism expressed by American, English, and European biologists, anthropologists, and psychologists prior to the Second World War, he actually examined the primary data on which they based their claims of the larger brains and superior minds of northern Europeans. In every case the samples had been deliberately biased, or the data misrepresented, or even invented, or the conclusions misstated. The consistently fraudulent data on IQ produced by Cyril Burt had already been exposed by Leo Kamin, but this might have been dismissed as unique pathology in an otherwise healthy body of inquiry. The evidence produced by Steve Gould of pervasive data cooking by an array of prominent investigators made it clear that Burt was not aberrant, but typical. It is widely agreed that ideological commitments may have an unconscious effect on the directions and conclusions of scientists. But generalized deliberate fraud in the interests of a social agenda? What more radical attack on the institutions of "objective" science could one imagine?

Being a radical in the sense that informs this memorial is not easy because it involves a constant questioning of the bases of claims and actions, not only of others, but also of our own. No one, not even Steve Gould, could claim to succeed in being consistently radical, but, as Rabbi Tarfon wrote, "It is not incumbent on us to succeed, but neither are we free to refrain from the struggle."

About the Contributors

NATALIE ANGIER, whose science writing for *The New York Times* won her the 1991 Pulitzer Prize, started her career as a founding staff member of *Discover* magazine, where her beat was biology. In 1990 she joined the *Times*, where she has covered genetics, evolutionary biology, medicine, and other subjects. Her work has appeared in a number of major publications and anthologies, and she is the author of three books: *Natural Obsessions,* about the world of cancer research (recently reissued in a new paperback edition); *The Beauty of the Beastly*; and the national bestseller, *Woman: An Intimate Geography,* published originally in 1999 and now available in paperback. She was the editor for Houghton Mifflin's *The Best American Science and Nature Writing 2002,* and she is currently working on a new book, *The Canon: What Scientists Wish That Everyone Knew About Science.* She is also the recipient of the American Association for the Advancement of Science–Westinghouse Award for excellence in science journalism and the Lewis Thomas Award for distinguished writing in the life sciences. She lives in Takoma Park, Maryland, with her husband, Rick Weiss, a science reporter for *The Washington Post,* and their daughter, Katherine Ida Weiss Angier.

"As children," she reports, "my younger brother and I sneaked in a middling amount of junk television. I say 'sneaked' because, should my father happen on us watching, say, *Gilligan's Island* or *The Flintstones,* he could match King Lear in howling rage, once going so far as to throw a heavy object against the television screen. Yet through that cracked screen our entire family each week watched the one show that my father loved: *Star Trek*—the original series, of course. I, too, adored all things *Enterprise*: Bones McCoy and his eye bags, Captain Kirk and his ever-ripping shirts, the cosmic love-ins, the beehive hairdos, the Star Fleet–issue miniskirts.

"I wrote my story about interstellar space travel as a kind of paean to *Star Trek*. Despite the seductive scenarios described in the story, I'm skeptical that we'll get very far in our space travels, and more doubtful still that we will ever encounter alien civilizations. The distances between stars are just too huge. Nevertheless, I can't help wishing that someday, one of my descendents will have cause to utter, in all seriousness, that magic command: 'Beam me up, Scotty. Beam me up now.' "

PETER CANBY, the head of fact-checking at *The New Yorker*, is the author of *The Heart of the Sky: Travels Among the Maya*. He has written articles and reviews relating to Latin America and the natural world for numerous publications. In addition to traveling extensively throughout the world, he has worked on a scallop dragger and built a solar-heated house. He lives in New York City.

"Since my trip through Nouabalé-Ndoki, much has changed," he says. "There is now not just a road to Makao; Makao has become a logging depot with Central African traders, prostitutes and wild price inflation. New sawmills ring the park and a capillary network of logging roads creeps ever closer. So far, however, thanks to outstanding cooperation between the Wildlife Conservation Society (the primary park administrator) and Congolaise Industrielle des Bois (the logging company that is cutting most of the land surrounding Nouabalé-Ndoki), the park itself has held up pretty well. But who knows how long that will last? Nouabalé-Ndoki's preservation is the result of a distressingly thin act of institutional faith.

"For this reason, several readers told me they found this piece depressing. I hadn't thought of it that way. Consider that studies of the oldest Nouabalé-Ndoki trees show a uniform age of a thousand or so years. Consider also that the park's sandy streams contain numerous small, black, petrified oil-palm nuts that have been carbon-dated to the same period. The fact that oil palms are not indigenous to the Nouabalé-Ndoki area and are a marker of West African civilization leads researchers to suspect that the Nouabalé-Ndoki river valleys were settled and then abandoned a thousand years ago. By whom is not at all clear. Everything changes. Everything is mysterious."

LEONARD CASSUTO is an associate professor of English at Fordham University. He is the author of *The Inhuman Race: The Racial Grotesque in American Literature and Culture* (1997), and the editor of two other volumes. Currently at work on a literary and cultural history of twentieth-century American crime fiction, he also writes frequently on academic politics. He lives in Washington, D.C.

"The scandal at Bell Labs," he writes, "is at least as much about the process by which scientific knowledge is endorsed and disseminated as it is about science itself. The story of the Jan Hendrik Schön affair initially intrigued me because I thought it might lead to a scientific version of the culture wars, with scientists coming under hostile scrutiny from groups that help finance them. It didn't turn out that way.

"The Beasley commission (whose report is available online at http://www.lucent.com/news_events/researchreview.html) found that Schön had committed scientific misconduct. Lucent Technologies immediately fired him and he remains disgraced. After a brief flurry of editorializing by the general press, any further outward ripples from the scandal were contained. Over a period of months following his dismissal, many of Schön's papers—not just the ones cited in the Beasley report—were formally retracted. The crisis passed.

"The Beasley report has provoked some salutary reform efforts by physicists. In particular, the committee stressed their discomfort with Schön's former mentor and supervisor Bertram Batlogg's role in the affair, even as they noted that he hadn't violated any existing guidelines for scientific misconduct. The report essentially called for new guidelines to be written to cover what Batlogg did—and what he didn't do. The American Physical Society duly approved a set of supplementary guidelines for coauthors and collaborators in late 2002, and at the same time endorsed a proposal for all aspiring physicists to take a course in scientific ethics. The American federal government, whose guidelines for scientific conduct remain the international standard, hasn't done anything yet. Batlogg himself reversed his previous course of denial and elusiveness after the Beasley report came out. He circulated an apology by email to many physicists, including those who had publicly criticized him. This effort at rehabilitation has not exactly succeeded for him; a vocal group in the physics community, led by Nobel laureate Robert Laughlin of Stanford, continues to hold Batlogg accountable for lack of scientific professionalism.

"There's also the matter of peer review. Most scholars in the humanities view the peer review system the same way that Winston Churchill viewed democracy: as the worst form of government, except for all the others. Physicists, on the other hand, trust the system as a way of separating good work from bad, and they want to protect it. Some of my sources wrote to me after Schön was fired to express relief that peer review had worked after all. But had it? It seems to me that Schön was exposed not by the formal peer review process, but by conscientious whistle-blowers in his profession. The inability of his fellow scientists to duplicate his results would presumably have come a cropper at some point, but that hadn't happened yet. Who knows how long it would have taken? And what about the next time?"

TREVOR CORSON began journalistic writing as a teenager traveling in Asia and was among the first generation of American students to attend college in the People's Republic of China. He went on to earn a B.A. from Princeton University in East Asian Studies and lived for several years in Buddhist temples in Japan. When he returned to the United States he moved to a small island off the Maine coast and worked as a commercial fisherman. Subsequently he was managing editor of *Transition,* a journal based at Harvard University; under his direction *Transition* won the Alternative Press Award for international reporting three times and was nominated for a National Magazine Award in general excellence. His articles and essays have appeared in *The Atlantic Monthly, The New York Times, The Los Angeles Times, The Boston Globe,* and other publications. His first book, to be published by HarperCollins next summer, will take up where the article in this volume leaves off. The book tells the story of an unusual collaboration between a community of Maine lobstermen and an eccentric group of scientists, detailing along the way shocking and often humorous revelations about the secret life of the American lobster.

"During the summers of my childhood," he explains, "I lived with my grandparents on Islesford, the Maine island described in 'Stalking the American Lobster.' As a teenager I yearned to work aboard a lobster boat. After attending college and spending five years in East Asia I returned to Islesford and became a 'sternman.' For the next two years I woke at four-thirty and worked ten- and twelve-hour days hauling up traps. It was a grueling routine, but an exciting time to be lobstering, because the catch in Maine was skyrocketing. While the scientists charged with managing the lobster resource warned of overfishing, I witnessed the homegrown conservation techniques the lobstermen practiced. When I traded in my fish-oily gloves for a pen and wrote the article, I learned more about lobsters than I imagined there was to know, but I learned a more basic lesson as well. Despite a childhood on Islesford my background is as far away from commercial fishing as you can get—my father was a scientist, and a staunch conservationist. The lobstermen of Islesford taught me that people who aren't scientists, but whose daily lives depend on an intimate understanding of the natural world, can be conservationists too."

JOSEPH D'AGNESE is a journalist and children's book author who lives in Hoboken, New Jersey. He is coauthor, with Nell Newman, of *The Newman's Own Organics Guide to a Good Life* (Villard/Random House).

He writes: "This story was suggested to me by a hematologist who happened to remark on the high costs associated with using chimps in research. Scientists I approached were unwilling to be interviewed about their biomedical work with primates, presumably because they feared exposure or reprisals from activists. Still hoping to put a human face on the issue, I visited the Fauna sanctuary. I happened to be there the day one of the chimps took sick and died. I came back and wrote about what I'd witnessed. In the end the faces at the center of the piece were not human at all."

THOMAS EISNER was born in Germany and grew up in Uruguay. He received his undergraduate and graduate degrees from Harvard, and has been a member of the Cornell faculty since 1957. An enthusiastic entomophile, he has written over four hundred papers on insects, their behavior, communicative skills, and survival strategies. A widely acclaimed photographer, and recipient of many scientific awards, he is an avid musician and dedicated conservationist. He has helped make award-winning film documentaries and received the National Medal of Science in 1994.

"I have long been fascinated by mosquitoes and their seeming ability to defy humanity on all fronts," he says. "To most people, mosquitoes are deserving of extinction—period. I thought they could stand a bit of positive publicity, so I decided to write about their sexual antics for *Wings,* the official magazine of the Xerces Society, the only organization dedicated to the preservation of invertebrates. I was always intrigued that it should have been an engineer, rather than a biologist, who discovered that mosquitoes get their high from a buzz as they buzz up high, and the story seemed worth telling, at least to the reasonably inclined."

ATUL GAWANDE has been a staff writer on medicine and science for *The New Yorker* since 1998 and recently completed his surgical training. He is now a general and endocrine surgeon on the staff of the Brigham and Women's Hospital in Boston and an assistant professor of health policy at the Harvard School of

Public Health. He is also the author of *Complications: A Surgeon's Notes on an Imperfect Science,* which was a finalist for the 2002 National Book Award for Nonfiction.

"The learning curve is something you think about from the very first day you put on a white coat," he explains, "and for good reason. It has terrors, important consequences, and vexing moral dilemmas. A perfect subject for an essay, I thought. I was nervous taking on the topic, though. There are only untidy solutions to the dilemmas. And no matter how carefully I explain why the opportunity to practice upon human beings is vital to good medicine, I (and many of my colleagues) feared the essay would just increase the number of people turning up in doctors' offices insisting that only the most experienced take care of them. But in truth, people have already figured out that experience matters. And offering an understanding of where it comes from and how seemed to me the only chance of leading anyone to accept the limits inherent in what we do and also our constant need to learn."

MARCELO GLEISER holds the Appleton Professorship of Natural Philosophy and is professor of physics and astronomy at Dartmouth College, where he leads an active research group in theoretical physics. To date, he has published over sixty-five papers in refereed journals and has participated in many domestic and international conferences as an invited speaker. He is the recipient of the Presidential Faculty Fellows Award (PFF) from the White House and the National Science Foundation and is a Fellow of the American Physical Society. His first book, *The Dancing Universe: From Creation Myths to the Big Bang* (Dutton, 1997), received the 1998 Jabuti Award, the highest literary award in Brazil. He has appeared in several science documentaries, including the PBS/BBC *Stephen Hawking's Universe.* He received the 2001 José Reis Award for the Popularization of Science, offered every two years by the Brazilian Research Council (CNPq). His second book, *The Prophet and the Astronomer: A Scientific Journey to the End of Time* (W. W. Norton, 2002), received the 2002 Jabuti Award. Since September 1997, he has written a widely popular weekly column in *Folha de São Paulo,* one of the top newspapers in his native Brazil.

He writes, "When Charles Harper invited me to contribute an essay to the volume celebrating Sir John Templeton's ninetieth birthday, I was elated. He suggested I write on the general topic of 'emergence' from the point of view of a physicist. Nothing could be more appropriate; the emergence of form from substance, be it of living matter from inorganic molecules, of mind from brain, or of the universe itself (from nothing?), is a topic at the forefront of scientific research. And it is also a very old question, much older than what we today call science. As such, it represents very uniquely the drive we all have to ask questions about Nature's mysteries and to try and answer them as best we can. This essay is an effort to communicate my own personal drive, a scientific drive fueled by a sense of awe which is also much older than science."

ROALD HOFFMANN was born in 1937 in Zloczow, Poland. Having survived the war, he came to the United States in 1949 and studied chemistry at Columbia and Harvard Universities. Since 1965 he has been at Cornell University, now as the Frank H.T. Rhodes Professor of Humane Letters. He has received many of the honors of his profession, including the 1981 Nobel Prize in Chemistry (shared with Kenichi Fukui). "Applied theoretical chemistry" is the way Roald Hoffmann likes to characterize the particular blend of computations stimulated by experiment and the construction of generalized models, of frameworks for understanding, that is his contribution to chemistry. Dr. Hoffmann is also a writer of essays, nonfiction, poems, and plays. The latest of his four poetry collections is *Soliton*, published in 2002. His nonfiction writing includes a unique art/science/literature collaboration with artist Vivian Torrence, *Chemistry Imagined: The Same and Not the Same*, a thoughtful account of the dualities that lie under the surface of chemistry; and, with Shira Leibowitz Schmidt, *Old Wine, New Flasks: Reflections on Science and Jewish Tradition*, a book of the intertwined voices of science and religion. Dr. Hoffmann is also the presenter of a television course, *The World of Chemistry*, aired on many PBS stations and abroad. A play, *Oxygen*, by Carl Djerassi and Roald Hoffmann premiered at the San Diego Repertory Theatre in 2001, and has had several productions since.

"This one was easy," he comments. "Have I not been peddling theories all my life? I should know what I preach.

"It was easy, but not for that reason. Scientists are mostly unreflective about what they do as they do it. Oh, they're very good at spotting lack of logic, obfuscation, and hype in other scientists. But not in their own work. And perhaps it's just as well—we all know too much thinking and talking about the process undermines creation. There is cognition and thought, mind working with hands, in the heat of making the new, yes. But not all that much stand-back-and-ponder-why thinking. At some point, it's just 'do it!'; as other theorists, I did what comes naturally. Does the reflective tone of this article then mean that I am through doing real science?

"I am not going to answer that question.

"I have been fortunate to have to rise to the occasion of writing *American Scientist* columns for a dozen years, alternating between popularized chemistry, chemical stories with a point, history or social issues, and amateur philosophy of science. 'Why Buy That Theory?' belongs to the last category. Michael Weisberg, a young philosopher of science and a friend, invited me to a symposium at the Philosophy of Science Association meeting in 2002, on the theme 'Causation and Explanation in Chemistry.' It was also high time for my next *American Scientist* column. I wrote 'Why Buy This Theory?' to . . . see where it would take me, as I had trouble beginning my talk. And because I was inclined to fight a little with all too rational ways of looking at science by philosophers and scientists.

"What may not be so obvious is the personal conflict (read: inconsistency) revealed in this article. First of all, the success of my early theoretical work with Woodward was based in substantial part on some risky predictions. Second, I have made a good living teaching people in chemistry simple orbital pictures of the driving forces for shape and reactivity. Respectful of complexity, I've still simplified—some would say oversimplified—the world.

"But in 'Why Buy This Theory?' I set off, bang, by dismissing the importance of risky predictions in theory acceptance. And I come out, desperately trying to restrain myself, for complexity.

"Why am I fighting myself? Is it that I've just gotten older? And as one ages one loses (some people do) the simple, strong convictions of the young? And sees shades of gray, the shadows that lurk around simple worldviews.

"No doubt that's part of it. But also that I've learned something from the ambiguity that gives a poem (or prose) meaning beyond simple meaning. That I just know more chemistry, more stories. And more people, who make wonderful molecules and build ornate theories, blissfully ignoring the Ockham's razor they idolize. People who give us the gift of new means of looking. Their way there is rife with tension, paved with inconsistencies as they craft provisional (all the while subtly claiming absolute) knowledge. Telling stories, not fessing up to it, telling them anyway, because they have. Just people, perforce fallible, relentlessly curious, driven to create the new."

JENNIFER KAHN writes about science and other subjects for *Discover, Harper's Magazine,* and *Wired* magazine, where she is also a contributing editor. She is based in Berkeley, California, and was recently awarded the American Academy of Neurology's 2003 journalism fellowship.

"A decade ago," she writes, "as an undergrad in the Princeton physics department, I remember seeing a crank letter pinned to the basement bulletin board. It was a long letter, written entirely in capitals and very neat, asking whether anyone knew about the government's ability to transmit radio messages through silver fillings. What struck me at the time was how reasonable the question was. Why couldn't fillings act like antennae at some frequency? I mean, how would *you* account for voices that seemed to originate inside your own head? Because I was in lab at the time, and struggling to explain the bizarre data that my experiments inevitably generated, I had a lot of sympathy for the idea that rogue electromagnetic waves permeated the universe. They had to be mostly undetectable, of course—but really, it would have explained a lot."

MICHAEL KLESIUS is a staff writer at *National Geographic* magazine, where he has spent the last ten years researching and writing science articles. He holds a master's degree from the Johns Hopkins science writing program in Washington, D.C. During his undergraduate years at the College of William and Mary, he excelled at languages and the written word, but found himself continually drawn to science courses and lectures for their mind-bending facts, theories, and controversies. During his junior year in France he crisscrossed much of the European continent and has returned to it a dozen times. On assignment for *National Geographic,* he has worked in China, Russia, South Africa, Syria, Thailand, Turkey, and Zambia. Haiti counts as one of his most rewarding stops, due to the limitless spirit of its people amid abject poverty. Topics he has covered for *National Geographic* include Neolithic cultures, the global AIDS pandemic, Iron Age ships excavated from Danish peat bogs, and new technologies in aviation, for which he flew aerobatics in the F-16 and F-18. Among his most memorable experiences, was trekking above Mount Everest's base camp to the peak of Nepal's Kala Pattar. Michael and his wife, Giuliana, live in Arlington, Virginia.

"Writing about science offers me a constant lesson in humility," he says, "both because the people I interview are orders of magnitude smarter than I am, and because I'm always left with the reminder of humanity's brevity and unremarkable place in the cosmos. I've always shared *National Geographic's* fascination with things ancient. So I eagerly accepted this assignment chronicling the rise of the angiosperms, or flowering plants. Reporting the story from Sweden to China to Wyoming's Big Horn Basin, I encountered paleobotanists as passionate about their calling as any scientists I've known. They showed me how flowering plants, extant and extinct, have played a critical role in the rise and sustenance of humans, and not just physiologically. As a flower dealer in the Netherlands said, 'People have been fascinated by flowers as long as we've existed. It's an emotional product. People are attracted to living things. Smell, sight, beauty are all combined in a flower. Every Monday a florist delivers fresh flowers to this office. It is a necessary luxury.' "

BRENDAN I. KOERNER is a contributing writer for *Mother Jones,* a contributing editor at *Wired,* and a fellow at the New America Foundation. He was formerly a senior editor at *U.S. News & World Report*, where he covered everything from paleontology to cybercrime. His work has appeared in *The New York Times Magazine, Harper's Magazine, Slate, The Washington Monthly, The New Republic, The Christian Science Monitor*, and *Legal Affairs*. He also writes the "Mr. Roboto" technology column for *The Village Voice*. Koerner was named one of the *Columbia Journalism Review*'s "Ten Young Writers on the Rise" in 2002, and he recently won the National Headliner Award for magazine feature writing. A 1996 graduate of Yale, he lives in New York City.

" 'Disorders Made to Order' was something of a mea culpa on my part," he writes. "As a novice journalist, I was assigned a story on the 'hidden epidemic' of social anxiety disorder, a malady with which I was not familiar. An editor assured me that it was a seriously underreported phenomenon, and that the Social Anxiety Disorder Coalition could point my reporting in the right direction. Indeed, the Coalition was only too willing to assist, eagerly providing interview subjects, scientific data, and enough colorful anecdotes to fill several magazines.

"Yet as I delved more deeply into the story, I began to sense the taint of drug-industry money. The Coalition's flacks doubled as press agents for SmithKline Beecham; the interviewees appeared in Paxil marketing videos; and the scientific talking heads were all paid consultants of one pharmaceutical giant or another. The kicker was the fact that the Coalition's creation was suspiciously timed to coincide with the Food and Drug Administration's approval of Paxil to treat SAD. Rather than write a heartfelt portrait of extremely shy souls who'd been helped by psychopharmacological treatment, I opted for a piece that focused on the drug industry's sly marketing tactics—and vowed to write a more in-depth account when I had the chance. It took three years to find a magazine willing to indulge my little quest.

"The trick was to write something more substantial than a cynical take on pharmaceutical flackery. Rather, we wanted to take a hard look at how the drug industry not only sells pills, but diseases as well. It can take upwards of a decade to discover and test a new drug; creating a new disorder, or expanding the patient base for an old one, is a far more cost effective process. Creepier, too."

For the last forty years RICHARD C. LEWONTIN and RICHARD LEVINS have worked in the same academic institutions, first at the University of Rochester, then the University of Chicago, and currently at Harvard University. Richard Lewontin is a population and evolutionary geneticist who has investigated the forces operating on genetic variation in natural populations, and works in the philosophy of science and in the political economy of agricultural research. He has been active in the radical science movement, including Science for the People and the Sociobiology Study Group. At present he is research professor in the Museum of Comparative Zoology at Harvard. Richard Levins is an ex-farmer turned ecologist. His primary interest is the study of processes in complex biosocial systems. He has worked in evolutionary ecology and population genetics with application to agriculture and public health, biomathematics, and philosophy of science and has been active in the New World Agriculture and Ecology Group and the New York Marxist School. At present he is a professor of population sciences at the Harvard School of Public Health and works at the Cuban Institute of Ecology and Systematics.

They write: "Over the last forty years we have worked both together and in parallel to attempt to create a coherent evolutionary population biology, with Dick Lewontin emphasizing the intricacies of genetic systems and Dick Levins focusing more on ecology. We have also worked to promote a critical science, aware of its insertion into the larger society, and a science politics that considers the organization, uses, political economy, recruitment, socialization and internal content of science. Our views on these matters are expressed in the essays contained in *The Dialectical Biologist.*

"In both the scientific and the political sides of our efforts we intersected with Steve Gould in many ways. He shared our view of the complex and historically contingent nature of living systems and their evolution, and he was a political ally. One or the other of us, or both, taught jointly with Steve in courses on evolution and on biology and society. We worked together with him in Science for the People and the Sociobiology Study Group, struggling against naïve biological determinism. All three of us shared a feeling of distance from many of our colleagues and from Harvard as an institution. It seemed appropriate, then, that the editor of the *Monthly Review* should ask us to write a joint memorial to him."

CHARLES C. MANN, the author or coauthor of four or six books ("depending on whether you count writing the text for books of photography," he says), is a correspondent for *Science* and *The Atlantic Monthly.*

"The genesis of '1491,' " he explains, "may lie in the day in 1984 when, writing an article about a NASA group that was monitoring the atmosphere, I landed with their specially equipped plane in Mérida, on the Yucatán Peninsula in Mexico. The atmospheric scientists had a day off in Mérida, and we all took a decrepit Volkswagen bus to Chichén Itzá. I knew nothing about Mesoamerican culture—somehow the true inventors of zero had been skipped in my math classes. But purely in aesthetic terms I thought these ruins were much more interesting than those I had seen in Europe. On my own—sometimes for vacation, sometimes on assignment—I went back to the Yucatán five or six times. For the German magazine *Geo,* photographer Peter Menzel and I made the fourteen-hour drive down a one-lane dirt road to the then-unexcavated city of Calakmul. We stayed in a *chiclero*'s shack in the midst of broken stelae. I still remember my amazement when our Maya guide, Juan de la Cruz Briceño, emerged from the forest with a wild turkey that he had caught by sneaking up to it and lopping its head off with his machete.

"In other words, the article stems from a long-standing though rather formless personal interest. This interest only snapped into anything resembling focus in September 1992, when by chance I saw a college library displaying the special Columbian quincentenary edition of the *Annals of the Association of American Geographers,* which contained Bill Denevan's manifesto, 'The Pristine Myth.' A year or two later, at the annual meeting of the American Association for the Advancement of Science, I attended a forum called something like 'The Genesis of the Amazonian Forest,' which featured William Balée of Tulane University and Anna C. Roosevelt of the Field Museum in Chicago. In his fascinating talk about anthropogenic forests, Bill Balée mentioned the explosive impact of Hank Dobyns, whose demographic work sounded interesting enough to send me back to the library. 'Gee,' I thought as I read *Their Number Become Thinned,* Dobyns's account of native American demography, 'someone ought to put this stuff together. It would make a really interesting article.' I kept waiting for that article to appear. The wait grew more frustrating when my son entered school and was taught the same things I had been taught—ideas that I knew had long been sharply questioned. Since nobody else appeared to be writing this article, I finally decided to take a stab at writing it myself. '1491' is the result of my efforts, part of a larger work in progress."

SUSAN MILIUS says she has learned to enunciate carefully when explaining that she writes about organismal biology on the staff at *Science News.* She remembers as a kid fantasizing about becoming a nineteenth-century plant explorer, but she ended up falling into journalism instead. After some initial bouncing around various niches (even working as magazine food editor for a while, although she freely admits she's a dreadful cook), she has focused on writing about biology for magazines, newspapers, and wire services.

"I fret about creeping fur-ism in the science press," she adds, "a subtle tendency to publish more stories about animals than about plants. I'm guilty of it myself because I find it hard to convey botanical excitements that compete with cute furry faces and weird habits of procreation. Thank heavens this project had plants doing things about as dramatically as they can, what with turning fire-engine red and hovering on the brink of death, and thank heavens especially for plant physiologists who're fun to talk to.

"I remember when molecular biology took over the world for a while during the last century, and if you wanted to fit in with the smart kids you admitted interest in ants and flowers and newts only as childhood memories. It's such fun now to see the borders blurring and the molecular people talking to the whole-organism people."

SIDDHARTHA MUKHERJEE was born in New Delhi, India, in 1970. He was graduated from Stanford University in 1993 with a degree in biology, and then completed his D.Phil. in Biological Sciences at the University of Oxford as a Rhodes scholar. He finished his M.D. at Harvard Medical School in 2000 and joined the Massachusetts General Hospital as a resident in internal medicine. He is currently a Clinical Fellow in Oncology at the Dana Farber Cancer Institute/Partners HealthCare System, a teaching affiliate of Harvard Medical School. His research focuses on the biology of cells in early development, including stem cells; his writing focuses on the intersection between science, medicine, and politics.

"As a graduate student in biology," he writes, "I was often asked what sort of research I 'worked on' in my laboratory. Answering that question—I discovered—meant making a devil's choice between the simplistic and the esoteric. On one hand, I had the option of a sweeping, glib and vague response: I worked on viruses that caused 'cancer,' I could say, stressing that incandescent word, and hoping that all conversation that followed would become illuminated in its glow. Cancer research, after all, was a scientific talisman, a sanctified area that floated singularly above doubt and derision, a field that no one could call abstruse or academic.

"But the other answer—and perhaps the more accurate one—was infinitely more detailed: as a matter of fact, I spent most of the time meticulously picking apart a specific protein, from a specific virus, that happened to be linked to cancer. By the time I had finished *that* story my audience's eyes would often glaze over with boredom. Yes, my research was linked to finding a cure for cancer—but it was only obtusely, distantly linked. I wasn't testing chemotherapies; I wasn't tracking rates of lung carcinoma among smokers. Between my laboratory bench and a cure for cancer there lay a long, long stretch.

"This tension—between curiosity-driven science and science driven purely by application—is felt by almost every biologist. And the question inherent in this piece—what is the right formula for finding a balance between these two goals?—is nearly unanswerable, partly because you couldn't easily design a study to answer it. In June 2000, while I wrestled with that question in my own career, I suddenly came upon Collier's example. As a medical student at Harvard, I remembered thinking of Collier as a quintessential example of a curiosity-driven scientist—and here, right before my eyes, he was becoming a champion of application-driven research. Vannevar Bush had envisioned this conversion nearly fifty years ago. And the prospect of this happening within the lifetime of a scientist was obviously heartening to a young biologist holed up in a laboratory with mice—who hoped to someday practice medicine among humans."

LIZA MUNDY is a staff writer for *The Washington Post Magazine.* She grew up in Roanoke, Virginia, and received her undergraduate degree from Princeton University and a master's in English literature from the University of Virginia. Before coming to the *Post* she worked at the Washington *City Paper,* where, among other topics, she wrote about the deaf pride movement at Gallaudet, the world's only university for the deaf. There and at the *Post,* she has also written about ethics and reproductive technology. She has freelanced for *Lingua Franca, The Washington Monthly, Redbook,* and *Slate.*

"It must be such a strange thing to be written about," she writes. "I think about this every time I report a magazine feature. For a private citizen, to agree to be profiled requires a leap of faith; a surrendering of control; a willingness to trust (and spend time with) a stranger whose job it is to ask intrusive questions; a laying-open of your life to the judgment of the reading public. This is true for anybody, but it was truer still for Sharon and Candy, who knew that their efforts to have a deaf child would attract opinion from many quarters. Sure enough, within days after the article appeared, their computer crashed from the number of emails they received from around the world. Some messages—from hearing and deaf alike—were critical. Many—from hearing and deaf alike—were galvanized, thoughtful, enlightened, supportive. Throughout, Sharon and Candy bore this scrutiny with graciousness and good humor, qualities that were severely tested when, shortly after his first birthday, their beloved son died, tragically and unexpectedly, of pulmonary hypertension, a congenital disease unrelated to his deafness. I hope that this story will now stand as a memorial to Gauvin. Precisely because his mothers were brave enough to be written about, and to let him be written about, his life touched more people—started more conversations, challenged more minds—than many that last much longer."

MICHELLE NIJHUIS is a contributing editor of *High Country News,* an environmental bimonthly that has covered the wonders and tragedies of the western United States for more than thirty years. Her freelance work has appeared in *The Christian Science Monitor, Audubon,* the online environmental magazine *Grist,* and other regional and national publications. In her life before journalism, she studied biology at Reed College in Portland, Oregon, and spent several years working as an itinerant frog and tortoise biologist in the Southwest and California. She's now settled in a small town in western Colorado, where she lives and writes in a straw-bale house built by her husband, Jack.

"After I wrote 'Shadow Creatures,' I became a magnet for urban wildlife stories," she says. "Every city dweller, it seemed, had spent at least some time shooting air guns at crows, or spying on coyotes, or yelling at sedentary Canada geese. One reader, who had watched javelinas graze on his landscaping in Tucson, marveled at their ability to swallow even his prickly-pear cacti. There was exasperation in most of these stories, but always more admiration than annoyance. These animals, after all, have figured out how to survive in modern human society—something every human knows is no small feat."

DANIELLE OFRI is the author of *Singular Intimacies: Becoming a Doctor at Bellevue* (Beacon, 2003), and is editor in chief and cofounder of the *Bellevue Literary Review.* Her stories have appeared in both medical and literary journals, as well as several anthologies. She is the recipient of the 2001 *Missouri Review* Editor's Prize for the essay *Merced,* which was selected by Stephen Jay Gould for *Best American Essays 2002.* She is also associate chief editor of the award-winning medical textbook, *The Bellevue Guide to Outpatient Medicine.* An attending physician at Bellevue Hospital and on the faculty of New York University School of Medicine, she lives in New York City with her husband and two children.

Of "Common Ground," she says: "If it is unsettling to write about the private affairs of a patient, it can be agonizing to write about one's own. For my patients, I am acutely cognizant of, and troubled by, the ethical issues and so take pains to alter names and identifying characteristics. For myself, this is not so easy. Before I'd written 'Common Ground,' I'd never shared this episode with anyone, save one close friend. Well into my late thirties, married, with children and several advanced degrees, I still hadn't told my parents about what had occurred when I was seventeen. When I'd initially written this essay, I created two versions: one with, and one without, my own experience contrasted to the patient's. I wrestled with which version to publish, then finally decided to include myself. As a writer, I realized that I have no choice but to seek the truth, and that overrode my personal queasiness and lingering doubts. And as a physician who brings the agonies and deepest vulnerabilities of her patients to paper, I can be no less brutal with myself. If my patients have their guts revealed to the world, how can I hide behind a white coat or a writer's pen?"

LAWRENCE OSBORNE was educated at Cambridge and Harvard. He is the author of a novel, *Ania Malina*; the travelogue, *Paris Dreambook*; a collection of essays about Catholicism, *The Poisoned Embrace*; and, most recently, a book about autism, *American Normal*. His *The Accidental Connoisseur* will be published by Farrar, Straus and Giroux in 2004. Osborne lives in New York City, where he is a regular contributor to *The New York Times Magazine, Salon,* and *The New York Observer.*

"I wanted to write this piece," he reports, "because genetics is the most glamorous frontier of contemporary science, but also the most fraught with anxiety. It seems that cloning, transgenic animals and genetic engineering are where our deepest nightmares and optimistic dreams come together. Yet, at the same time, there may well be a pragmatic invention that binds them together very simply. Such seemed to be the case with spider-goat silk: human ingenuity at its most quietly daring. To me, it was irresistible."

DENNIS OVERBYE is a science correspondent for *The New York Times.* Born in Seattle, he majored in physics at MIT before deciding that writing was the only thing he was fit for. His first job in journalism was as a part-time assistant typesetter at *Sky and Telescope* magazine. His articles have appeared in a variety of publications and he is a two-time winner of the American Institute of Physics Science Writing Award. He is the author of *Lonely Hearts of the Cosmos: The Scientific Search for the Secret of the Universe* (HarperCollins, 1991), which was a finalist for the National Book Critics Circle Award, and *Einstein in Love: A Scientific Romance* (Viking, 2000). He lives in Manhattan with his wife, Nancy Wartik, and their daughter, Mira Kamille.

"My article," he says, "was written on the occasion of Stephen Hawking's sixtieth birthday, which was celebrated with a weeklong series of scientific talks and parties in January 2002. It was a moving event, full of traditional English fare such as butlers, toasts, and Marilyn impersonators. Hawking is arguably the most famous and most recognizable living scientist, St. George in a wheelchair battling the black-hole dragon to the millions of readers of his books. That he had a career at all and made it to the age of sixty after being diagnosed with Lou Gehrig's disease back in his twenties is amazing, a testament to his grit and the divine whims of providence. Scientifically, his greatest legacy, besides his students, is likely to be his discovery in 1973 that black holes are not really black and will in fact eventually explode due to quantum effects. A quarter of a century later that discovery is still reverberating; the tale of its adventures is an example of how the meaning of a powerful insight can change and grow with time."

GUNJAN SINHA was born in Bihar, India, but grew up in Brooklyn, New York. She earned a graduate degree from the University of Glasgow, Scotland, in molecular genetics. Her first job was in an organ transplant lab, which turned out to be surprisingly dull. After months of careful thought, she decided to try her hand at journalism and graduated from New York University's Science and Environmental Reporting Program in 1996. She has been writing about science ever since. She was life sciences editor of *Popular Science* for five years and also wrote about general science and technology. In 2000, she was awarded the Ray Bruner Science Writing Award that honors reporters who demonstrate exceptional ability early in their career. She is now a freelance journalist in Frankfurt, Germany, where she is often seen whizzing around on her bicycle.

"One Valentine's Day," she writes, "a friend sent me a story from *The Boston Globe* she thought I'd find 'amusing.' We'd just had a long conversation about relationships and office crushes and the story was a short, fun, fluffy piece about the biochemistry of love. I thought it would make a great feature. I looked into it some more and was fascinated by the progress scientists were making on the subject. 'Isn't this cool?' I thought. 'For millennia, love has been the domain of poets and philosophers. But now, for the first time in history, scientists are coming up with their own biological understanding of love.' But of course, there's much more to the potent biochemical stew that makes and breaks lives. Scientists continue to study fine distinctions of the emotion, such as how the love between mother and child differs from the love between husband and wife, for example. For me, the most interesting aspect of all this research is that it suggests there's another reason people screw around, one that has nothing to do with the love you didn't get as a child or other analysis of your psyche. The behavior is hardwired—a primal drive that's perfectly natural, even if it is hugely destructive."

FLOYD SKLOOT is the author of nine books. His most recent is a memoir of living with brain damage, *In the Shadow of Memory* (University of Nebraska Press), which was a Barnes & Noble Discover Great New Writers selection for summer 2003. In spring 2005, Louisiana State University Press will publish his fourth collection of poetry, *The End of Dreams*. Skloot's work has been included in *The Best American Essays 1993* and *2000, The Best American Science Writing 2000, The Best Spiritual Writing 2001,* and *The Art of the Essay 1999.* He lives in rural western Oregon with his wife, Beverly Hallberg.

" 'The Melody Lingers On' is from my book-in-progress, *Fragmentary Blue,*" he comments. "It's a memoir counterpointing the relentless destruction of my mother's memory with the slow reassembling of fragments of my own memory, which was shattered by brain damage following a viral illness. In the year and a half since 'The Melody Lingers On' was written, my mother's condition has continued to worsen, but song remains. There are fewer songs left now, and their snatches of lyric or melody are even less coherent, but on occasion her face will still light up and her voice will fill the room for a moment."

MARGARET WERTHEIM is the author of *Pythagoras' Trousers,* a history of the relationship between physics and religion, and *The Pearly Gates of Cyberspace: A History of Space from Dante to the Internet.* Originally from Australia and now living in Los Angeles, her articles have appeared in many magazines and newspapers, including *The New York Times, The Sciences, New Scientist, The Times Literary Supplement, The Guardian,* and *Salon.* She writes the "Quark Soup" column for *LA Weekly* and is a regular contributor to *The Los Angeles Times Book Review.* She has also written a dozen television science documentaries, including the PBS special *Faith and Reason* and the award-winning Australian series *Catalyst,* which was aimed at teenage girls. She and her husband have just completed a new documentary, *It's Jim's World . . . We Just Live in It,* on visionary outsider physicist James Carter. She is currently working on a new book, *Imagining the World,* about the role of imagination in theoretical physics.

Of "Here There Be Dragons," she writes: "This piece has a rather strange history. I was taken to the Mirror Lab at the University of Arizona by Father George Coyne, the Jesuit priest and astronomer who is head of the Vatican Observatory. I was writing a profile of Coyne for *Wired* magazine and one day he suggested a visit to Roger Angel's lab, where the mirror for the Vatican Advanced Technology Telescope was cast. Although *Wired* published the piece, they had just run a long story on telescope technology and decided to cut the section on the Mirror Lab. But it's such a superb location and Angel has such an incredible story that I felt the material had to be used somewhere, so I rewrote it for my 'Quark Soup' column. It was one of those lovely serendipitous ideas that came out of left field and then ended up somewhere very unexpected. Sometimes pieces take on a life of their own, and I feel that's very much what happened with this one."

FRANK WILCZEK is a theoretical physicist aspiring to become a natural philosopher. He is currently the Herman Feshbach professor of physics at MIT. He's won prizes for both science and writing, including recently the Lorentz Medal of the Netherlands Academy and the Lilienfeld Prize of the American Physical Society. He is a product of the New York City public schools.

"A lot of my best work deals with the basic mathematical laws that govern the interactions of elementary particles," he writes. "It's not easy to explain this kind of material in an honest way to anyone who lacks either extensive training or unusual patience. But I'm often asked to give public lectures, or to write for a general audience. And I think this is an important thing to do, since I believe frontier science is a most valuable and beautiful production of our culture, one that ought to be widely shared. Wrestling with this challenge, I had the happy thought that the important message of my work is not this or that arcane fact, but that Nature, though She speaks an unfamiliar language, is not only comprehensible but brilliantly logical. That's what this essay tries to convey."

Acknowledgments

A Note from the Series Editor

Submission for next year's volume can be sent to:

Jesse Cohen
c/o Editor
The Best American Science Writing 2004
Ecco/HarperCollins
10 E. 53rd St.
New York, NY 10022

Please include a brief cover letter; manuscripts will not be returned. Submissions made electronically are also welcomed and can be e-mailed to <jesseicohen@netscape.net>.

ecco

Thought-Provoking and Exciting Scientific Inquiry

An annual series dedicated to collecting the best science writing of the year from the most prominent thinkers and focusing on the most current topics in science today.

THE BEST AMERICAN SCIENCE WRITING 2003
ISBN 0-06-093651-7 (paperback)
Edited by Oliver Sacks
Oliver Sacks, one of the foremost thinkers/writers on neurology and medicine, and the bestselling author of *The Man Who Mistook His Wife for a Hat* and *Awakenings,* edits this fourth installment of the annual series. Featuring articles from: Peter Canby, Charles C. Mann, Atul Gawande, Liza Mundy, Floyd Skloot, Frank Wilczek, Marcelo Gleiser, Natalie Angier, Margaret Wertheim, Jennifer Kahn, Michelle Nijhuis, Gunjan Sinha, Trevor Corson, Siddhartha Mukherjee, Michael Klesius, Susan Milius, Thomas Eisner, Lawrence Osborne, Brendan I. Koerner, Joseph D'Agnese, Danielle Ofri, Roald Hoffmann, Leonard Cassuto, Dennis Overbye, and Richard C. Lewontin & Richard Levins.

THE BEST AMERICAN SCIENCE WRITING 2002
ISBN 0-06-093650-9 (paperback)
Edited by Matt Ridley
Edited by the renowned and bestselling author of *Genome*, Matt Ridley, the third volume in the series includes pieces by: Lauren Slater, Atul Gawande, Lisa Belkin, Margaret Talbot, Sally Satel, Jerome Groopman, Gary Taubes, Joseph D'Agnese, Christopher Dickey, Michael Specter, Mary Rogan, Sarah Blaffer, Natalie Angier, Julian Dibbell, Carolyn Meinel, David Berlinksi, Tim Folger, Oliver Morton, Steven Weinberg, Nicholas Wade, and Darcy Frey.

THE BEST AMERICAN SCIENCE WRITING 2001
ISBN 0-06-093648-7 (paperback)
Edited by Timothy Ferris
Pulitzer Prize and National Book Award nominee Timothy Ferris, one of the preeminent writers about astronomy, edited this second volume of outstanding science writing. The contributors include: John Updike, Michael S. Turner, Natalie Angier, Joel Achenbach, Erik Asphaug, John Archibald Wheeler, Stephen S. Hall, Richard Preston, Peter Boyer, John Terborgh, James Schwartz, Ernst Mayr, Greg Critser, Andrew Sullivan, Malcolm Gladwell, Helen Epstein, Debbie Bookchin and Jim Schumacher, Stephen Jay Gould, Tracy Kidder, Jacques Leslie, Robert L. Park, Alan Lightman, and Freeman Dyson.

THE BEST AMERICAN SCIENCE WRITING 2000
ISBN 0-06-095736-0 (paperback)
Edited by James Gleick
This first volume in an annual series carries the imprimatur of Pulitzer Prize nominee James Gleick, the celebrated chronicler of scientific social history. This stellar collection includes the writings of James Gleick, George Johnson, Jonathan Weiner, Sheryl Gay Stolberg, Deborah M. Gordon, Francis Halzen, Timothy Ferris, Stephen S. Hall, Floyd Skloot, Denis G. Pelli, Douglas R. Hofstadter, *The Onion*, Don Asher, Natalie Angier, Stephen Jay Gould, Susan McCarthy, Peter Galison, and Steven Weinberg.